Shock and Awe in Fort Worth

Also by Sheryl Elam Tappan

Proposal Secrets: How Bechtel and Other Big Firms Really Win Those Billion-Dollar Government Contracts (In Preparation)

Shock and Awe in Fort Worth

How the U.S. Army Rigged the "Free and Open Competition" to Replace Halliburton's Sole-Source Oil Field Contract in Iraq

Sheryl Elam Tappan

Pourquoi Press • San Mateo, California

SHOCK AND AWE IN FORT WORTH:
How the U.S. Army Rigged the "Free and Open Competition" to Replace Halliburton's Sole-Source Oil Field Contract in Iraq

Published by Pourquoi Press
P.O. Box 651
San Mateo, CA 94401-651
(PourquoiPress@msn.com)

ISBN 0-9749947-8-2

1 2 3 4 5 6 7 8 9

Library of Congress Control Number: 2004091107

Printed in the United States of America
Distributed by BookRegime.com

For the unsung heroes —
the thousands of honest, hard-working employees
of Halliburton, Bechtel, and other contractors
who left families and lives of safety at home
to help reconstruct Iraq —
and whose dedication and accomplishments,
working under horrendous conditions, have been
overshadowed by the focus on "friends in high places"

CONTENTS

ABOUT THE AUTHOR

Sheryl Elam Tappan, an independent consultant, has written proposals that have won billions of dollars' worth of U.S. Government contracts. In early 2003, she was responsible for the proposal that won Bechtel the first Iraq civil infrastructure reconstruction contract from the U.S. Agency for International Development (USAID), a contract valued at $1.3 billion. In the summer of 2003, she led the Bechtel proposal team in the Army's competition for Iraq oil infrastructure reconstruction work — until she discovered the competition was a sham and recommended the company withdraw, which it did. In the fall of 2003, she wrote much of the proposal that won another $1.8 billion worth of Iraq work from USAID.

Sheryl Tappan's other winning work for Bechtel includes proposals for operating the Kwajalein missile range in the Pacific for the Army's Strategic Missile Defense Command, a $4.5 billion contract, and for managing up to $5 billion worth of Cooperative Threat Reduction projects in the former Soviet Union for the Defense Threat Reduction Agency. These projects range from eliminating missiles and missile bases in Russia and Ukraine to securing biological weapons and production facilities in Kazakhstan and Uzbekistan. She has assisted other engineering, construction, and program management companies in improving both the content of their proposals and the process they use to prepare them (TappanConsulting@msn.com).

Sheryl Tappan earned a B.A. in Zoology from the University of California at Berkeley and an M.A. in Marine Biology from San Francisco State University. After some years of working for aquaculture and environmental consulting companies, she devoted nearly 10 years to writing — or managing teams of authors writing — proposals for large government contracts, including six years exclusively for Bechtel. She lives near San Francisco with her husband, James G. Tappan, M.D., an obstetrician/gynecologist, and they have two sons, Matthew (hers) and Paul (his). Ms. Tappan is a steadfast Democrat, and her husband a staunch Republican. His father, Paul W. Tappan, J.D., Ph.D., was a frequent visitor to the Eisenhower White House; he headed the commission that wrote the U.S. Penal Code and chaired the Federal Parole Board.

ACKNOWLEDGEMENTS

As the book's author, I accept full responsibility for all of its faults, while sharing any praise it may receive with a number of friends, associates, and perfect strangers. I am grateful to my dear friend, Susan Margolis, and to Peter Ginsberg of Curtis Brown Ltd., Bernard Goodwin of Prentice Hall, and Dan Poynter of Para Publishing for their advice and encouragement. Many thanks to the journalists cited in this book for their invaluable reporting and to Craig Albert for the impetus to write it. Most of all, my eternal gratitude belongs to my husband, Jim, for his contributions to my work and my life.

DISCLAIMER

This book was written to provide information and opinion on the U.S. Government contracts mentioned herein, and on associated actions and activities conducted by U.S. Government employees or others, based on the personal experience and perspective of the author. Although every effort has been made to ensure that all information presented in this book is true and accurate, it is sold with the understanding that the publisher and author are not legal experts and are not privy to all information pertaining to the contracts, actions, and activities described. There may be mistakes, both typographical and in content; therefore, this book should be used only for general guidance and opinion, not as the ultimate source of facts. Furthermore, this book contains information that is current only up to the printing date. The publisher and author shall have neither liability nor responsibility to any person or entity with respect to any loss or damaged caused, or alleged to have been caused, directly or indirectly, by information contained in this book. If you do not wish to be bound by the above, you may return this book to the publisher for a full refund.

PREFACE

Every American must have known by the summer of 2003 that Vice President Richard Cheney once ran oil giant Halliburton and that Halliburton was the center of the firestorm surrounding the award of Iraq reconstruction contracts to political cronies of the Bush administration. If Americans read the papers or watched the news, they could not help but believe that Bush and Cheney had been doling out multi-billion-dollar contracts to Halliburton and Bechtel because the corporations had made million-dollar contributions to their political campaigns or, in the case of Cheney and Halliburton, because there was some other financial gain. Were Bechtel and Halliburton simply handed billions of dollars' worth of work in Iraq by "friends in high places," as the administration's critics wanted us to believe, or were the contracts actually awarded fair and square in full compliance with the law?

Big business is Republican, but that is not enough to explain the success of big companies like Halliburton and Bechtel. During the Clinton administration, Halliburton subsidiary Kellogg Brown & Root (KBR) was awarded a sole-source contract to support the U.S. peacekeeping mission in the Balkans, but that did not make the news. No one accused Democrats Clinton and Gore of involving our country in a war in Eastern Europe to benefit Halliburton, a major contributor to the Republican Party. That would have been absurd. During the Clinton administration, Bechtel won billions of dollars' worth of U.S. Government work, but no one accused Bechtel, another major contributor to the Republican Party, or board member George Schultz, Secretary of State under President Reagan, of eliciting political favors from the Clinton administration. That would have been a *non sequitur*, too.

The Bush administration's critics are predominantly liberal Democrats, which suggests political motivation on their part as well. In April 2003, they accused Schultz of calling in a political favor when Bechtel won the first Iraq civil infrastructure reconstruction contract from the U.S. Agency for International Development (USAID) over six other large companies invited to compete. USAID is part of the Department of State, but who would Schultz have called? Who, in a position of power at State, still owed Schultz a favor after 15 years? And how would that person have achieved the desired goal, given the large bureaucracy involved in the laborious procurement process mandated by federal law? The firewalls that separate USAID's political appointees from the career civil servants that conduct competitive procurements make awarding a large contract to a political patron of a high-ranking official extremely difficult.

None of the critics who bombarded the public with accusations of political cronyism throughout 2003 had any proof that either Halliburton or Bechtel "bought" their Iraq contracts. They had not identified the bureaucratic pathway by which the political payback could have been accomplished or even a single link in the chain. Nevertheless, whether one voted for Clinton and Gore (as the author did) or Cheney and Bush, Republicans begetting contracts to Republicans was easy enough to understand, so the belief in political payback persisted. Political cronyism and corruption make good sound bites, so the media were content to report the same accusations over and over without uncovering any hard evidence at all.

The media seemed to believe, like most Americans, that a president or a vice president had the power to order a billion-dollar contract award with no bureaucratic scrutiny whatsoever, but this is a myth if it is not based in fact. The media's "facts" came from corporate watchdog organizations that alleged political corruption based, not on proof, but on circumstantial evidence in the form of misleading campaign contribution statistics and sophistic, overly simplistic interpretations thereof. Similar "facts" came from the top Democrat on the House Government Reform Committee, who painted a picture of waste and gold-plating that he claimed was enriching Halliburton and Bechtel at U.S. taxpayers' expense.

But no one advanced so much as a theory as to the bureaucratic route by which the political cronyism could have been achieved. They did not have the hard evidence they needed to support the circumstantial charges they made. They apparently did not know where to look or the right questions to ask, perhaps because they did not understand how the federal procurement process really works. No wonder. The processes and procedures the U.S. Government uses to select and oversee contractors are virtually impenetrable for anyone who does not deal with them on a daily basis. Even the best investigative journalist would have a hard time untangling the details. So the accusations of political favoritism persisted because it was what most people assumed to be true.

But automatically equating a lucrative federal contract with political payback does a disservice to our nation, for it does not produce the nonpartisan understanding needed if the real problems in U.S. Government contracting are to be identified and resolved. The federal procurement process is flawed and inefficient in many ways, and it is occasionally corrupt but not as often as people think. When political cronyism does occur, "the devil is in the details." That's where the media, the watchdogs, and the House Government Reform Committee should have gone to find the proof they needed that procurement fraud had occurred, including the names of the conspiratorial links in the bureaucratic chain from Cheney to Halliburton KBR. No one did that or made it public until now.

The U.S. Government's dependence on private industry has been increasing at a dramatic rate, due in part to the downsizing that began under Clinton and Gore. Between 1990 and 2001, the Defense Department's acquisitions workforce — the personnel who select and oversee private contractors — was cut by more than 50%, while the number of contracts that had to be awarded and administered increased by over 12 percent. And it's only going to get worse. The Pentagon is expecting more than half of its remaining civilian acquisitions staff to retire by 2007, and the Bush administration has implemented a plan to outsource hundreds of thousands of additional federal jobs. Every outsourcing contract does, of course, add to the workload of the remaining acquisitions staff.

Since September 11, 2001, the contracting challenges facing the Pentagon and other federal agencies have been greater than ever. With no end in sight to the war on terrorism, there will be no relief soon for the thousands of military and civilian contracting officers and staff in Washington, Baghdad, and other hot spots around the world. Contractors cannot be sent into situations like those in Iraq when the war ended using fixed-price contracts; private industry cannot be expected to assume all the risks in performing the government's work. Fixed-price (lump sum) and fixed-unit-price contracts can only be used when the government's requirements are well defined; unknowns lead to change orders, cost overruns, and other potential nightmares for both the contractor and the contracting officer. Cost-reimbursable contracts are fully justified in situations like post-war Iraq, where the security threats and work requirements changed constantly; cost-reimbursable contracts were awarded to Halliburton and Bechtel to reconstruct the Iraq oil fields and civil infrastructure, respectively. Cost-reimbursable contracts require much more time and expertise on the part of the military personnel or civil servants entrusted with contractor selection and contract oversight. With less time and expertise available in the federal workforce, there will undoubtedly be greater and greater potential for procurement fraud, contract mismanagement, and other abuses, all at the taxpayer's expense.

By the end of 2003, many books had already been published about the war in Iraq. Political bestsellers had attacked the Bush administration from the left, and others had defended it from the right. Until now, no book has examined the firestorm surrounding Halliburton's Iraq contracts, especially from the inside of the military/industrial complex where the author sits. In the author's opinion, the U.S. Army Corps of Engineers is guilty of contracting abuses, ranging from gross mismanagement to procurement fraud, above and beyond the contracting abuses attributable to Halliburton KBR. The facts presented in this book demonstrate unequivocally how the Corps of Engineers arrogantly manipulated and unnecessarily delayed the Iraq oil field contracting, from start to finish, in order for KBR to get virtually all the work. The facts, discovered in documents published by the Corps, show how Corps personnel lied to the other bidders in the supposedly "full and open competition" for a significant share of the work. The facts show how Corps personnel hid valuable information from the other bidders, information that Halliburton had and that the other bidders were entitled to as well.

This was the "full and open competition" required by the Competition in Contracting Act and promised in March 2003, when the Defense Department was forced to reveal that it had secretly awarded KBR a sole-source contract. The Pentagon claimed the contract was only temporary; it would enable emergency work to be performed while a competition was held. However, the contract had a ceiling in the stratosphere — $7 billion — way beyond any legitimate definition of temporary or short-term. The competition turned out to be a sham, but the contractors who knew it have not complained for fear of reprisal and hurting their chances for future federal contract awards.

The Corps personnel identified in this book are critical links in the bureaucratic chain that connects Vice President Cheney to Halliburton KBR's Iraq contract awards. Whether the highest-ranking link identified in this book, a U.S. Army general, connects directly to Cheney is yet to be determined; perhaps there were one or two additional links in between. Nevertheless, the author steadfastly believes that political cronyism is the exception, not the rule, in federal contracting. When it does occur, its perpetrators must be routed out; they cannot be allowed to go uncensored. We cannot allow the myth to be perpetuated that political cronyism is business as usual in any administration. Major reforms in government contracting <u>are</u> desperately needed for the long road ahead, especially in overseeing cost-reimbursable contracts, whose necessity will not go away. We need to start here and now to get our government's contracting house in better working order.

Sheryl Elam Tappan
San Francisco
March 1, 2004

1

Total Devastation: Where There's Chaos, There's Opportunity

AS FAR AS THE EYE CAN SEE in every direction, the world is black. Stinging from the noxious gases that fill the air, the eyes squint, but there is no horizon, no demarcation between the blackness of the sky and the blackness of the land, no demarcation between heaven and hell. High noon is as dark as a moonless midnight, and it rains sticky drops of oil non-stop. The black rain spreads from Eastern Europe to Pakistan. Black smoke billows hundreds of feet into the air from 650 oil wells burning out of control. For months they burn, emitting 2.5 billion cubic feet of smoke that carries 4 million tons of sulfur and soot that obliterates the sky. The lighter fractions vaporize into the air as noxious chemicals that form thick, acrid clouds, while the heavier fractions fall to the earth, covering the desert with tar. The tar hardens into a thick layer of sludge that persists for years.

Another 100 well heads detonated by Saddam's fleeing troops are gushing oil, 50 million barrels that create 200 lakes where once there was land. The oil oozes and seeps down through the sand, aiming for the groundwater below. Five million more barrels of oil float on the Gulf, 600 times more than the Exxon Valdez released in Alaska in 1989. Nearly 30,000 birds are dead, their bodies washed up on the shore. It is the worst man-made environmental disaster in the history of the world. Most of the refineries and pipelines are also destroyed. Where fields of fuel storage tanks once stood, only melted hulks of metal remain. Over 85 percent of the country's oil production infrastructure must be rebuilt. Experts say it will take two to three years to extinguish the oil fires alone. Meanwhile, the black rain persists.

It will take tens of thousands of workers to complete the massive oil field restoration and reconstruction to return production capacity to what it was before the war. But where will they sleep? Where will they eat? Compounding the challenges is the total lack of facilities to support tens of thousands of workers in the desert. There are no hotels, no hospitals, no food, no nothing. Nor is there water to fight the fires; the nearest source may be tens or hundreds of miles from the wells. Everything must be brought in, everything from portable housing and heavy construction equipment to fireproof underwear for the crews battling the 2000 degree blazes of the 650 oil fires. But before that can happen, other crews, unexploded ordnance specialists, must clear the way, for Saddam's troops left thousands of hidden land mines and other dangerous debris in their wake.

This is the scenario the Bush administration expected to find in post-war Iraq because it is exactly what occurred in Kuwait after the first Gulf War. Experts were convinced that Saddam Hussein would destroy his own country's oil fields in a final act of defiance, and their restoration would cost tens of billions of dollars. So in the fall of 2002, the Defense Department began planning for the certainty of the devastation just described times two, for Iraq had twice as many wells as Kuwait. The following March 6, 2003, two weeks before the war began, the Pentagon announced that it had "reliable reports" that Saddam had 24 rail cars full of pentolite explosives to use in his final act of terrorism against the Iraqi people, Iraq's neighbors, and the world. The Pentagon announced that it had developed a plan for fighting the oil well fires and, on the morning of March 13, posted a toll-free telephone number on its Web site. Companies that specialized in extinguishing oil well fires could call and be added to the bidders list. The companies knew the opportunities in Iraq would be even greater than they had been after the Gulf War — Iraq's 1,500 oil wells were spread over an area 20 times larger than Kuwait. By afternoon, the Pentagon's voicemail box was full.

2

Fighting Fire with Fire: The Iraq Oil Field Contingency Plan

THE DECISION TO GO TO WAR is alleged to have been made by the Bush administration in early fall 2003, if not long before. That fall, the Pentagon began the planning required to deal with the hundreds of oil well fires and massive damage they expected to find in the oil fields in Iraq, which is to say the Defense Department assigned a contractor the responsibility for developing an Iraq oil field contingency plan. Due to the uncertainty of when the war would start and when it would end, the Pentagon did what was most expedient. It assigned the task to Halliburton KBR under an existing multi-task-order contract known as LOGCAP, short for Logistics Civil Augmentation Program. Won in 2001, KBR's current 10-year LOGCAP contract with the Army Materiel Command's Operations Support Command "essentially turns KBR into a kind of for-profit Ministry of Public Works for the military."[1]

The Pentagon has argued justifiably that it did not have the many months the government's procurement process requires to conduct a free and open, or even limited, competition if the winning contractor was going to do the planning and be on the ground in Iraq ready to fight the fires when the conflict ended. If a new contract had been competitively bid, it would logically have included both the planning and execution phases for the sake of expedience.

Assigning the contingency planning to Halliburton KBR under LOGCAP was contractually allowed and expedient, but was it wise? The Pentagon had two other choices. The Federal Acquisition Regulation (FAR) allow for a sole-source award, i.e., a contract awarded without competition, to the most highly qualified contractor in the case of "unusual and compelling urgency." Even without time constraints, the FAR allow for a sole-source award to a contractor with unique qualifications. Both of these options could be exercised quickly and yield better results if there was indeed a contractor more highly or uniquely qualified than Halliburton KBR. Of course, the contractor would have to be willing and able to rapidly take the assignment on. For a contract awarded without competition, the FAR require the Defense Department to prepare a formal sole-source justification document that describes the market research conducted to determine the most highly or uniquely qualified contractor and the results of that research. If no market research was conducted, the justification must adequately explain why not.

Under the LOGCAP contract, Halliburton KBR's tasks had ranged from building temporary bases and fixing trucks to peeling potatoes and cleaning latrines. Specialized hazardous work, such as demining a construction site, was subcontracted out. Firefighting oil fires in Iraq would be very specialized and very hazardous work. It would be work like that in Kuwait after the first Gulf War. The oil field restoration and reconstruction program in Kuwait was the nearest precedent to the massive program of firefighting and rebuilding that the Pentagon thought would be needed in Iraq when the conflict ended, given the intelligence on the last-ditch plans of Saddam Hussein.

Even the briefest of market research would have produced only one result. There had never been another project in the history of the world, except the one after the Gulf War in Kuwait, that had anywhere near the same combination of specialized work scope, occupational safety hazards, and extraordinary logistical challenges, challenges far more demanding than any experienced under LOGCAP.

The Kuwaitis had hired a large American company to plan and execute the entire program, from extinguishing fires, capping wells, and cleaning up a sea of oil in the desert to rebuilding refineries, repairing pipelines, and reconstructing other war-torn facilities. The contractor received and screened over 30,000 unsolicited proposals for how to put the fires out quickly, such as constructing a giant dome over the wells and sucking out the air; without oxygen, the fires would go out.

Instead, the contractor hired specialized oil-firefighting subcontractors, including the legendary Red Adair, to comprise 28 firefighting teams, and the contractor found a way to transport 20 million gallons of water through the desert to the teams each day. The contractor built 200 giant reservoirs, each holding a million gallons, and filled them with seawater pumped from the Gulf through 90 miles of oil flow lines that Saddam's troops had installed, intending to dump Kuwait's wealth in the Gulf. The lines had to be cut, pieced together, and redirected to the shortest route between the Gulf and the fires; the shortest route was the only route that had been cleared of land mines, booby traps, and other unexploded ordnance debris. Heavy equipment was moved from site to site at night. It really didn't matter; it was so dark in the daytime that drivers often couldn't see their turns.

Nevertheless, the contractor's team extinguished the 650 oil fires burning out of control and capped the 100 wells gushing oil onto the land in only nine months. Various experts predicted it would take two or three, some even said five, years. To deliver firefighting water, they used oil pipelines and reversed the flow, saving $700 million in pipeline construction costs alone. Experts had estimated $4 billion as the total cost of extinguishing the fires and repairing the wells in order to get the crude oil flowing again. The project was called Al-Awda in Arabic, which means "the coming back of something there before." Al-Awda's final cost was only $2.3 billion. Edward "Skip" Gnehm Jr., the American Ambassador in Kuwait at the time, described the American contractor's achievement as "nothing short of amazing." The same American company was hired to manage the subsequent rebuilding of the damaged refineries and other oil production infrastructure in order to return Kuwait's oil production to the prewar level of 1.6 million barrels a day.

The Pentagon had made dire predictions of the damage Saddam Hussein's troops would do in the oil fields of Iraq, at least as much destruction as they left behind in Kuwait. So, if any market research was conducted, it would have found the most qualified — uniquely qualified, given the project's grand scale — contractor to do the Defense Department's planning, and later execute the plan. It was clearly the large American company the Kuwaitis had used, and that contractor was Bechtel, not Halliburton KBR. A Halliburton press release claimed that its crews had been responsible for extinguishing 320 oil well fires in Kuwait after the Gulf War, but that was blatantly untrue. Bechtel was hired by the Kuwaiti Oil Company as the general contractor and placed in charge of all of the firefighting and other oil field restoration and reconstruction work. Halliburton KBR was not one of the oil fire subcontractors that Bechtel hired to do the hands-on work.

It was not Halliburton KBR but Bechtel that met the mega-challenges after the first Gulf War. Kuwait had no manual workforce, so the contractor sent over 1,000 engineers and construction managers to Kuwait and recruited 10,000 laborers of 35 nationalities in the first six months. After the Iraqi troops withdrew, the Kuwaiti oil fields had no housing, food, running water, electricity, phone service, or heavy construction equipment left, so Bechtel conducted the largest peace-time airlift since Berlin. In June 1948, Stalin ordered Russian troops to surround the city, cutting off all overland means of supply. It was the beginning of the Cold War. Until the blockage was lifted in September 1949, a U.S.-led airlift of fuel and food kept 2.5 million people from freezing and starving to death. Experts had originally said it couldn't be done. Bechtel procured $730 million worth of equipment and supplies from a dozen countries, including the entire 1991 output of Caterpillar D-8N backhoes. A fleet of wide-body jets made 200 flights to deliver 200,000 tons of bulldozers, cranes, and other heavy equipment to Kuwait. The Russian-made Antonov, the world's largest cargo plane, was used to deliver long-reach backhoes and 150-ton cranes because it was the only plane large enough to load and offload the equipment without disassembling it. Bechtel also built and operated portable housing, a 40-bed field hospital, and mess halls serving 30,000 meals a day, as well as docking facilities, warehouses, motor pools, and communications systems.

The Pentagon continues to claim that no contractor was as qualified as Halliburton KBR, but Bechtel had demonstrated superior capabilities, having done in Kuwait exactly what the Pentagon thought would be required in Iraq. Bechtel had done it on the largest scale the world has ever known, and Bechtel's performance in Kuwait had been outstanding, which should have made it the contractor of choice. With the world watching the U.S. Government's every move for misuse of Iraqi oil funds, why trust the success of such an important mission to a contractor found to have significantly overcharged the government on LOGCAP tasks in the Balkans and other recent Defense Department work?

The Pentagon continues to claim that no other contractor could have been assigned the contingency planning and execution tasks as fast, but the Defense Department could have had Bechtel's services just as quickly and done it in any one of three ways. Bechtel would have jumped at the opportunity and brought valuable lessons learned on the ground in Kuwait to plan for the firefighting and oil field repairs in Iraq. The Pentagon could have justified a sole-source award based on Bechtel's unique experience in Kuwait, or it could have assigned the contractor the work through at least two existing U.S. Government contracts.

Bechtel is one of five large American engineering and construction companies awarded a Cooperative Threat Reduction Integrating Contract (CTRIC) by the Defense Threat Reduction Agency (DTRA), part of the Defense Department, after a full and open competition in 2001. The other four CTRIC contractors are Halliburton KBR, Parsons, Raytheon Technical Services, and The Washington Group (WGI). CTRIC is the same type of multi-task-order contract as LOGCAP in that, at contract award, both the quantity of work that will be assigned and when it will be needed during the 5-year contract period are unknown (indefinite delivery/indefinite quantity, or ID/IQ, in government terms). The five companies compete for individual task orders for U.S.-funded CTR projects in the former Soviet Union, or they are awarded sole-source task orders when justified according to the FAR. The projects include the elimination of intercontinental ballistic missiles, missile bases, and chemical, biological, and nuclear weapons production facilities and the interim safe storage of these weapons of mass destruction (WMD) until they can be eliminated.

Until Iraq, the Defense Department had not used the CTRIC contracts as a vehicle to procure services required outside the CTR program in Russia, Ukraine, Kazakhstan, and other now independent states of the former Soviet Union. But in December 2002, DTRA issued a request for a "white paper" from each of the five CTRIC contractors. "White paper" is the term used in government and industry for a document written for the purpose of presenting an agency or company's approach to an issue or a problem. White papers can be solicited by decision-makers or submitted unsolicited by one entity desiring to influence another. The DTRA-solicited white papers were to describe the CTRIC contractors' capabilities and approaches to eliminating the chemical, biological, and nuclear WMD the government was certain would be found in Iraq. DTRA paid $50,000 for each of the four papers submitted. On March 7, 2003, DTRA issued a request for proposal to the five CTRIC contractors, giving them only one week to respond, and by the end of the month DTRA had awarded two Iraq WMD disposition task orders, one to Halliburton KBR and one to Raytheon. Neither company has deployed a WMD disposition team to Iraq, for no WMD have been found, but the task orders prove that the Pentagon could have easily used the CTRIC contract to award a sole-source task order to Bechtel to plan the firefighting effort in Iraq. The DTRA WMD task-order competition also proved the Defense Department is capable of conducting a contract competition in less than a month.

The second way the Pentagon could have accessed Bechtel's services as quickly as Halliburton KBR's was through Bechtel's contract with the U.S. Department of Energy to manage and operate the Nevada Test Site. In doing so, Bechtel provides services to all of the site's customers, including the Defense Department. The 1,375-square-mile site in the desert 65 miles northwest of Las Vegas includes the National Counter-Terrorism Training Center and the Hazardous Materials Spill Center, where the latest technologies are developed for the U.S. Government and where military personnel and other first-responders from federal, state, and local agencies around the country come together for tabletop training and field exercises. The contingency planning for post-war Iraq included both firefighting and oil spill response, so taking advantage of the expertise at the Nevada Test Site would have been wise.

Instead, in November 2002, the Pentagon gave the task to Halliburton KBR under its LOGCAP contract with the Army. How could the Pentagon have thought Halliburton KBR was the best qualified contractor? Jumping forward to March 24, 2003, yields a clue. On that date, the Pentagon revealed that it had signed a contract on March 8 with Halliburton KBR to implement the contingency plan that KBR had prepared. The sole-source contract had been awarded secretly and without competition and so it caught the attention of Representative Henry A. Waxman (D-Ca.), Ranking Minority Member on the House Government Reform Committee, and corporate watchdog organizations like The Center for Public Integrity. It was the contract that sparked the media frenzy over political cronyism, because every American knew by then that Vice President Cheney had run Halliburton until election year 2000. When the sole-source award was announced publicly on March 24, 2003, a Defense Department spokesperson told the press that Halliburton KBR was the obvious choice for the sole-source contract because of its experience extinguishing 320 oil well fires in Kuwait after the first Gulf War!

How did the Defense Department come to believe that Halliburton was responsible for extinguishing nearly half of the 650 oil well fires in Kuwait? Presumably, Halliburton made the same false claim to the Defense Department that the company included in a press release on March 24: "In 1991, Halliburton crews brought 320 wells in Kuwait under control in less time than was expected following the Gulf War."[2] In an interview published in the leading engineering and construction industry journal, *Engineering News-Record,* on April 7, 2003, Terry Farley, the now-retired Bechtel executive who managed the contract with the Kuwait Oil Company, said Halliburton "put out no fires — zero."[3] Farley, a legion in the industry, won *Engineering News-Record's* Man of the Year Award in 1992.

Was no due diligence done? Neither Halliburton KBR nor Bechtel is an oil firefighting company. Both use subcontractors for the front-line, hands-on work, as Bechtel did in Kuwait. Given the magnitude of the conflagration they expected to find as Saddam's troops retreated through the oil fields of Iraq, what the Pentagon needed was a contractor to plan for the whole gamut of firefighting, spill response, and logistical support activities that would be required and, for greatest efficiency, to implement the activities when the war ended.

The Pentagon could have gone through DTRA to conduct a fast-track limited competition between the CTRIC contractors, including Halliburton KBR and Bechtel; all five CTRIC contractors are cleared for classified Defense Department work. The Pentagon could have conducted a fast-track limited competition as the U.S. Agency for International Development did for the Iraq civil infrastructure reconstruction contract and as allowed by the FAR. In either case, the Pentagon could have given the contractors only one week to respond as DTRA had done for the WMD. The Pentagon had choices.

The Pentagon has maintained that Halliburton KBR was the logical choice because it was already supporting preparations for the war in Iraq under its LOGCAP contract. The Army created the contract in order to have a private contractor on call to provide logistics support whenever and wherever needed, from constructing temporary military bases to doing laundry and delivering the mail. Outsourcing these more mundane aspects of a mission frees military personnel to concentrate on the mission and eliminates the need to activate reservists. When the Army requires support under the contract, it issues a specific task order, also known as a delivery order, which specifies the services needed. Thus such an open-ended contract is called a "multi-task-order," as well as ID/IQ earlier defined. This type of contract provides speed and flexibility to meet the military's rapidly changing needs in hot spots like Iraq; the LOGCAP contractor must deploy the required resources within 72 hours. Such contracts usually have maximum dollar amounts as well as guaranteed minimums, usually $500,000; however, KBR's current LOGCAP contract has no ceiling.

Back in 1991, the Department of Defense had hired KBR to write a top secret plan describing how private companies could provide the military with logistics support for missions anywhere in the world. The company was paid $3.9 million for the classified document and another $5 million to expand the plan a few months later.[4] In 1992, near the end of Dick Cheney's term as Secretary of Defense but before he had any known association with the company (Cheney became Halliburton CEO in 1995), KBR was awarded a contract by the U.S. Army Corps of Engineers to implement the plan the company had prepared. During the five-year term of the newly created LOGCAP contract, Halliburton KBR provided logistics support for U.S. military operations in Somalia, Rwanda, Haiti, and elsewhere. The largest assignment, worth $546 million, was in the Balkans.

Halliburton KBR lost the LOGCAP I contract to DynCorp at the end of its five-year term; DynCorp underbid KBR when the contract came up for rebid in 1997. When DynCorp's five-year LOGCAP II contract was up, Halliburton KBR won it back in the rebid in 2001, and the LOGCAP III contract term was increased to ten years. As of January 22, 2004, KBR had received LOGCAP III task orders totaling $3.6 billion.[5]

After Halliburton KBR lost LOGCAP I in 1997, the Army Corps of Engineers awarded the company a sole-source contract, worth $1 billion over two years, to continue supporting the U.S. peacekeeping mission in the Balkans. When that contract expired, KBR beat out one competitor for a five-year contract extension valued at $180 million a year. But when war broke out in Kosovo, the contract value skyrocketed to $1 billion in 1999 alone.

The lack of competition in the sole-source award in 1997, which occurred during the Clinton administration, did not make the news, for there was nothing newsworthy about a nonpartisan decision. According to the Pentagon, it made little sense to change contractors in midstream, especially if they were performing well. Case closed.

But not quite. The U.S. Government often changes contractors in midstream because a contractor may, for a variety of reasons, lose the rebid on a contract that supports a long-term endeavor. When contracts for services are awarded, they have maximum periods of performance, such as a three-year base period with two one-year options to extend, which the government exercises if all goes well. But apparently all did not go well for Halliburton KBR on its previous LOGCAP I contract. The government alleged that it had failed to deliver the services requested, or severely overcharged for them, on four of the seven task orders issued during the Balkans conflict. The General Accounting Office (GAO) auditors found "no evidence that cost was taken into consideration."

A GAO report in February 1997 had shown, for example, that KBR had overcharged the Army for plywood by billing it at $85.98 per sheet, more than six times the commercial price of $14.06.[1] The company claimed the price was high because it had to fly the plywood in from the States. So perhaps the sole-source award to Halliburton KBR for continued support in the Balkans later that year should have raised some eyebrows for lack of due diligence, just as the subsequent LOGCAP III contract award in 2001 should have. According to a GAO report in 2000, Halliburton KBR had also overcharged the government by providing more personnel, goods, and services than required, such as by cleaning Army offices four times a day.

Government auditors scolded the Army for "losing control of its contractor" and said the Corps "should have done more to control costs."[6] They said Army personnel "frequently have simply accepted the level of services the contractor provided without questioning whether they could be provided more efficiently or less frequently and at lower cost." The GAO report concluded, "Poor oversight by the military had let Halliburton's firms sell the Army unneeded services."

If the Army Corps of Engineers conducted due diligence in 2001 before awarding Halliburton KBR its current LOGCAP contract, the company would have had to have resolved the performance problems that occurred under its previous one. At least one safeguard was imposed on the new LOGCAP contract. Where Halliburton KBR's LOGCAP managers once had purchasing authority up to $500,000, they must now have an Army officer sign all purchase orders over $2,500. But Halliburton KBR had other trouble with the Army.

In 2002, the company paid $2 million to settle federal fraud charges arising from its accounting practices on a contract with the Army Corps of Engineers for work at Fort Ord, California, during the period when Cheney was Halliburton's CEO.[1] A former employee had blown the whistle and filed a lawsuit in 1997, alleging the company had inflated the costs of building materials, maintenance, and repairs, increasing its profits at the government's and the taxpayers' expense. He alleged that his supervisors at Halliburton KBR had instructed him to bill for work never performed and that the company had submitted false claims and made false statements related to over 200 task orders between 1994 and 1998.[7] Both the Defense Department and the Justice Department investigated before the matter was settled. According to T.C. McIntosh, a Defense Criminal Investigative Service agent involved in the case, the U.S. Attorney's office in Sacramento, California, dropped the case because of poor contract monitoring on the part of the Army Corps of Engineers.[8]

A series of class-action lawsuits for damages caused by exposure to asbestos-containing products manufactured by some of Halliburton's other subsidiaries led to a $4 billion settlement, also in 2002. Enron and other corporate accounting scandals were making the news that year, and Halliburton garnered its share. It was being investigated by the Securities and Exchange Commission for accounting improprieties during the years Cheney was at the helm. The company allegedly counted cost overruns as additional revenue on engineering and construction jobs before the customers had agreed to pay; the overrun costs totaled $100 million in 1998. Arthur Anderson was the company's auditor at the time.

Between the base period and options, the current LOGCAP contract could extend for 10 years. Halliburton KBR has received over two dozen task orders under LOGCAP to support the Defense Department in Iraq. The contract has no cost ceiling, and Halliburton KBR is reimbursed for all of its costs and paid an additional sum that equals up to 3 percent of the costs (it was 9 percent on the first LOGCAP contract). Many of the tasks orders are valued at over $60 million each, yet they were doled out as if they were menial jobs. There is no competition for LOGCAP work, as there is for CTRIC work where DTRA has five CTRIC contractors competing and keeping the costs in check through the force of competition. With so many hot spots around the world, why is the Pentagon putting "all of its eggs in one basket" with only one LOGCAP contractor? If Halliburton KBR were to become stretched thin, what would happen then to U.S. military troops?

By early April 2003, KBR had received 24 LOGCAP task orders totaling over $425 million to support Operation Iraqi Freedom.[7] In addition to the contingency planning, the task orders included $60 million for logistics support in Turkey and $69.5 million to support the Office of Reconstruction and Humanitarian Assistance led by retired General Jay Garner, which would be replaced by the Coalition Provisional Authority led by Ambassador Paul Bremer on May 6.

So there have been two known cases of the Defense Department hiring Halliburton KBR to prepare classified documents, the military logistics outsourcing report and the Iraq oil field contingency plan. The outsourcing report was prepared when Halliburton KBR was just beginning its first LOGCAP contract. The Iraq contingency plan was prepared as a task order under Halliburton KBR's current LOGCAP contract, and it led to a sole-source contract to implement the work it had planned. Both documents were classified, yet neither affected our national security, and all three noncompetitive awards occurred when Cheney was in charge, either as Secretary of Defense or Vice President. Despite denials by Cheney of any financial interest in Halliburton since he came to office this time, he left Halliburton in 2000 with stock options worth $33 million, and he continues to receive deferred salary payments of between $100,000 and $1 million a year. So the circumstantial evidence for political cronyism between Cheney and Halliburton KBR goes back to 1991-1992.

3

The Sole-Source Award: Halliburton's Coup de Grâce

BEFORE THE WAR EVEN BEGAN, the Defense Department had awarded Halliburton KBR its second sole-source Iraq assignment. On March 8, 2003, the company was awarded a contract to implement the very contingency plan it had prepared in November 2002. The contracting agency within the Defense Department was with the U.S. Army Corps of Engineers through its Southwestern Division, Fort Worth District. Although the award was made on March 8, it was not disclosed publicly until March 24. At that time, the Army said the contract was short-term and low-cost because the scope only included extinguishing well fires, assessing the damage to pipelines and other critical oil production facilities, and performing emergency repairs. But it would also include operating Iraqi oil production facilities and distributing fuel.

With the disclosure of the sole-source award, the media, corporate watchdogs, and Representative Waxman began leveling charges of political cronyism, which have intensified ever since. It would be weeks before Representative Waxman and others could force the Army to release the facts: Halliburton KBR's sole-source contract for supposedly temporary services actually had a two-year duration and a $7 billion ceiling! It was cost-reimbursable plus fees totaling seven percent of the costs, more than twice the LOGCAP contract fees, which totaled three percent.

Given such a large contract dollar value, the Army should have conducted a full and open competition, or at least a limited competition as DTRA did under CTRIC for the WMD disposition in Iraq and USAID did for the Iraq civil infrastructure reconstruction. DTRA and USAID completed their competitions in three months or less, and the Pentagon could have done the same. The Pentagon could have conducted a fast-track, full and open competition for one contractor to do both the contingency planning and implementation. Competition would have saved the government money, because contractors competing for cost-reimbursable contracts usually lower the percentage of contract costs that they want the government to pay as the fixed fee.

In response to a March 26 letter from Representative Waxman, the Army's Chief of Engineers, Lieutenant General Robert Flowers, wrote on April 8, "There will be ample opportunity for competitions of the overall requirements to support the restoration of Iraq's oil infrastructure... Only Kellogg Brown & Root Services [Halliburton KBR], the contractor that developed the complex, classified contingency plans, could commence implementing them on extremely short notice...Under the circumstances, no other contractor could satisfy mission requirements in the time available."[9]

Yet when Bush declared the war over on May 1, 2003, only nine oil wells had been set afire, and despite all of the contingency planning, procurement, and preparation completed by Halliburton KBR, it and its two firefighting subcontractors did little of the work. According to the *Wall Street Journal*, volunteers from the Kuwaiti Oil Company, firefighters trained by the Bechtel team after the Gulf War, did most of the work.[10] In late March, even before the giant southern Iraqi oil fields were under U.S. and British control, the Kuwaitis had extinguished the first oil well fire. In an area littered with minefields and Iraqi guerrillas, the Kuwaitis cleared several live antipersonnel cluster bombs themselves and completed the firefighting and well-capping job, while Halliburton KBR's crews were still in Kuwait City waiting for their equipment to be delivered and attempting to locate suppliers of firefighting water. The Kuwaitis were not paid, so why did they do it? According to the chief firefighter for the Kuwait Oil Company, it was a gesture of friendship to the Iraqi people — and to protect Iraq's underground oil reservoir, which overlaps with their own. The Kuwaitis put out the last oil fire on April 13; Boots & Coots had extinguished two others. The other seven oil fires set by Saddam burned themselves out.[11]

As more details of the sole-source contract became known, the extent of Halliburton KBR's coup de grâce was revealed. The contract was not only open-ended; its scope was far greater than extinguishing oil fires, assessing damage, and making emergency repairs. On May 2, in response to another letter from Representative Waxman, the Army's General Flowers disclosed that the contract also included major oil infrastructure reconstruction work. This was the work he had said other contractors would have ample opportunity to bid for in his letter of April 8.[9] General Flowers also stated, "This contract was designed from the outset as a bridge to competition and is structured accordingly. The indefinite delivery/indefinite quantity, cost plus award fee contract allows the Government to issue individual task orders for a specified period of performance at an estimated not-to-exceed dollar amount. We will limit orders under this contract to those services required prior to the availability of competitively awarded contracts."

On May 2, General Flowers wrote, "The government will limit orders under this contract to those services necessary to support the mission in the near term. The procurement strategy is to compete subsequent effort at the earliest reasonable opportunity consistent with the needs of the mission."[9] By May 20, Halliburton KBR had been issued $185 million worth of work under the sole-source contract.

It took nearly two months for General Flowers to respond to Representative Waxman's next letter, dated April 16. On June 10, the general stated, "There has been confusion in the press about the estimated costs of the current KBR contract and the new acquisition being planned. The cost of the current contract will be the sum of the cost of the orders placed under it. As of May 20, 2003, five orders have been placed. The total estimated cost of those five orders as of

May 27, 2003, is $184,786,000. We are limiting orders under this contract to only those services that must be performed or initiated prior to the availability of competitively awarded contracts… The estimated cost of future contracts is not known because the full extent of the work required on the Iraqi oil infrastructure is yet to be determined."[9]

On June 10, General Flowers wrote a second letter, this one in response to Representative Waxman's letter of May 6. He stated, "Preparation for and execution of this mission have been a challenge for the Department of Defense, since the condition of the oil infrastructure after military operations was an unknown and the only comparable situation was after the Kuwaiti war of 1991. In that war, sabotage by Iraqi troops damaged more than 700 oil wells and resulted in tremendous environmental damage and economic disaster in Kuwait. The Defense Department's planning had to consider and encompass the wide range of activities that might be necessary to handle a similar or possibly worse situation." [9]

It is hard to refute The New York Times' assessment of how Halliburton KBR "won" the contract without having to compete for it. "KBR didn't need any help," Dan Baum wrote on June 22, "It is by now so enmeshed with the Pentagon that it was able essentially to assign the contract to itself."[1] The Army maintained that Halliburton KBR got the contract because it was the only company that could meet the requirements outlined in the classified Iraq oil field contingency plan — never mind that Bechtel was the only company that had ever executed such a plan, when it did so in Kuwait after the Gulf War. If there had been a full and open competition for the contract, other bidders would have asked the Army to disqualify Halliburton KBR on the grounds of "organizational conflict of interest." Because KBR had written the plan, it clearly had an unfair competitive advantage. According to the FAR, an organizational conflict of interest exists when a contractor has information relevant to a contract up for bid that is not available to all competitors and when such information would assist in obtaining the contract award. Not only did the Army not conduct a competition, but it used what would have been KBR's conflict of interest to justify the sole-source award, when Army spokesman Lieutenant Colonel Gene Pawlik said, "They were the company best positioned to execute the oil-field work because of their involvement in the planning."[1] Colonel Pawlik's statement is, in effect, an admission that information in the classified contingency plan would have given Halliburton KBR a competitive advantage. He said KBR was "best positioned" to execute the work; he did not say KBR was best qualified.

In a June 10, 2003, letter to Representative Waxman, General Flowers said, "We are also nearing completion of our required acquisition strategy for the follow-on, competitively awarded contracts, if needed... the competitively awarded contracts will be one of many sources available to the Iraqi management team; the Iraqis will not be required to make use of the contracts and may procure required services from other sources if they so choose."[9] The Pentagon was beginning to hedge. "If needed" was a dramatic turnabout from General Flowers' May 2 pledge.

Perhaps General Flowers' two letters dated June 10 were prompted by one from Representative Waxman dated June 6, which began, "I write to request additional information on reports that the Army Corps of Engineers may drop its plan to replace the secret, no-bid contract awarded to Halliburton for oil work in Iraq with a contract issued under a competitive bidding process. If accurate, this would contradict numerous previous statements by the Corps, including assurances in letters that you have written me."[9] Representative Waxman's letter was prompted by a Dow Jones Newswires report on June 4, "U.S. May Drop Plan for Open-Bid Iraqi Oil-Field Contract."[12] Representative Waxman continued, "If the news accounts are correct, it appears that the Army Corps of Engineers has done an about-face and is poised to give additional benefits to Halliburton under its no-bid contract. Gary Loew, the planning director of the Corps's project, reportedly cited production deadlines and said, 'There may not be time to actually award a second contract.'"

As the media frenzy over political cronyism intensified in May and June, the Pentagon had flip-flopped, first announcing that the full-scale Iraq oil infrastructure reconstruction would indeed be conducted under a subsequent contract, awarded with full and open competition, and that the new contract would be in place by August. The value of that contract would depend on how much damage was done to Iraq's oil infrastructure during the war, how much damage Saddam's troops did during their retreat, including how many wells they set on fire, and how much damage has accumulated from other wars, sanctions, and decades of neglect. By some estimates, the Pentagon's Restore Iraqi Oil (RIO) program, as it was called, could take up to ten years and $50 billion. Task Force RIO was established to plan and coordinate the work. It consisted of representatives from the Iraq Ministry of Oil, the Coalition Provisional Authority under Ambassador Paul Bremer, and the U.S. Army Corps of Engineers. Task Force RIO was led by Brigadier General Robert Crear, who had commanded the Corps' Southwestern Division before the current assignment in Iraq.

With the announcement that a competitive procurement was planned, the media shifted its attention in the coming months to the fact that Halliburton KBR's price to deliver gasoline to the U.S. military and others in Iraq was nearly three times the cost of delivery by the Iraq Ministry of Oil. So there was less notice when in June the Defense Department did an about-face again, backing off on that time estimate and even stating that there might not be a new contract if the RIO mission were completed before a new contractor could take over the work or if the Iraq Ministry of Oil made their own arrangements to get it done. Asked about the August date, Colonel Pawlik said, "We're not going to try to discuss a specific timetable... I would be very surprised if that would be in the timetable, with all the requirements that are out there."[13]

4

The Two New Contracts: "Free and Open Competition" or Fraud?

AFTER REPEATED DELAYS, the U.S. Army Corps of Engineers posted a presolicitation notice on the Federal Business Opportunities Web site on June 23, 2003.[14] The notice said the Corps planned to award one or two contracts for "repair and continuity operations of Iraq oil infrastructure" to replace Halliburton KBR's. The contract or contracts would be open-ended and cost-reimbursable plus fees like the sole-source one. The period of performance would be two years plus three one-year options, and the ceiling would be $500 million. The prime contract competition would be open to all U.S. companies and companies in Coalition countries and Iraq. It would also be open to companies in countries qualifying under the Trade Agreements Act and Balance of Payments Act, which included NAFTA and Caribbean Basin nations. Companies in other foreign countries could be subcontractors to the two prime contractors. A bidders meeting would be held in Dallas on July 14, and the proposals would be due exactly one month later.

A few days after the presolicitation notice was posted, the Army posted a change, stating there would be two contracts for sure, one for work in the oil fields in the north of Iraq and one for work in the south. Awarding two follow-on contracts would be the perfect solution to the pressures on the Army for an open competition. One contract could be awarded to a new contractor, which would appease the political cronyism attacks, and the other could be awarded to Halliburton KBR to justify the sole-source award after the fact. That would be easy enough, now that the company had more RIO program knowledge and experience than any other firm.

The actual solicitation, more commonly known as the request for proposal (RFP), was posted on July 7.[15] U.S. Government RFPs are always organized the same way. The scope of work for a new contract is always Section C, but the Corps placed it in Section J, which is normally reserved for various attachments, such as plans prepared by the government or previous contractors that are germane to the new contract. In response to a question at the bidders meeting the following week, the Army said it had been a simple mistake and corrected it on July 21 in Amendment 001 to the RFP.[15] In hindsight, however, the Corps's quirky placement of the scope of work in Section J is telling, especially when some of the document's contents are examined. The work was divided into thirteen tasks, more or less chronologically, with Task 9 "reserved," rather than the last task, Task 13, suggesting the undefined work was in actuality known. Curiously, Task 8 was "internal distribution and sales of refined product."

Under each of the twelve defined main tasks, there were three sections: key subtasks, deliverables or work products, and procurement considerations. It was the third section that was strange because the scope of work in an RFP is supposed to list the new contractors' tasks, but this one listed responsibilities and concerns that are the government's, including the procurement of the two new contractors. How could the new contractors procure themselves? Under Task 1, for example, which is "establish management infrastructure" for the Iraq oil production facilities, there are two procurement considerations. The first one, "the work needs to begin ASAP to prepare for responsive execution," would go without saying. The second consideration, "not explicitly covered in Contractor plan or advance funding to date," is indeed strange if the document was what it was supposed to be, a scope of work prepared by Army procurement personnel for two new contracts. Who was the "Contractor" and to what plan and advance funding did the consideration refer? They could only be Halliburton KBR, the oil field contingency plan, and the hundreds of millions of dollars committed by then to the company to implement the plan under the sole-source contract, strongly suggesting that the document was prepared by Halliburton KBR, not the Corps of Engineers. Government procurement personnel would not have included their own concerns and job considerations in the new contractor's scope of work.

The language in Task 1 was subtle, but it became clearer in Task 2. Here, the procurement considerations included "establish initial facility operational control," where it said, "this and other work tasks should be stated broadly enough to provide for related contractor work as necessary to support the mission." Task 10, "repair oil system," included the most bizarre procurement consideration of all — "consider awarding this and subsequent tasks under competitive procurements"! Why was that in the new contractors' scope of work, when it was the Corps's role? Another Task 10 consideration said, "this is a major item — potentially over $1 billion." A similar consideration was listed under Task 12, "repair product systems (gas, refining)" — "this is a major item — potentially worth billions of dollars." Clearly, these procurement considerations were advice from Halliburton KBR to the Army. The title of the document should have been Iraqi Oil Infrastructure Restoration Scope of Work. Instead, it was entitled Iraqi Oil Infrastructure Restoration Contracting Strategy. The Corps's procurement personnel could have at least changed the title and edited out the procurement considerations before including the document as the new contractors' scope of work in the RFP posted for the world to see. Were they not concerned about the transparency of the charade, or were they just careless or incompetent?

5

A Theatrical Farce: The Bidders Meeting in Dallas

THERE WERE 199 PEOPLE in the hotel conference room that hot Texas day, July 14, 2003. There were 184 company representatives, presumably all bidders, either large prime contractors or smaller firms interested in joining a large team. In addition, there were twelve U.S. Army Corps of Engineers personnel, including Colonel Michael Schultz, acting Southwestern Division commander since Brigadier General Crear took over Task Force RIO in Baghdad. There were also three members of the press. All of the major players were present, including Bechtel, Fluor, and Foster Wheeler with three representatives each, Parsons with four, WGI with two, and Raytheon with one. Halliburton KBR had the largest contingent with thirteen representatives, including Charles ("Stoney") Cox, the company's RIO program director, who had flown in from Baghdad for the event, or perhaps to cool off. Other bidders came from as far away as Australia, China, and Korea, as well as the United Kingdom, Italy, and the Middle East. They came in the belief that there was a billion dollars' worth of work to be awarded, openly and fairly this time, to two winners to be announced on October 15. They believed it because that is what the Army's RFP said and because the Army's secret sole-source award to Halliburton in March had focused so much attention on the competitive procurement that the Corps would not dare do anything other than follow the FAR.

An understanding of what transpired at the bidders meeting, and how fraudulent it was, requires an examination first of a document the Corps quietly posted on the Web before leaving work on Friday evening, August 1, nearly three weeks later. The document, entitled *Restoration of Iraqi Oil Infrastructure Final Work Plan*, was dated July 24, 2003.[15] As stated in the document, it was the result of the joint collective effort of representatives of the Iraq Ministry of Oil and its associated organizations, the U.S. Army Corps of Engineers' Task Force RIO, Halliburton KBR, and the Coalition Provisional Authority. It said the joint effort had occurred at a workshop held in Baghdad July 6 through 9, 2003, the week before the bidders meeting in Dallas.

Stoney Cox knew the contents of the plan when he appeared at the meeting on July 14, and we can assume the twelve other Halliburton KBR attendees did as well. No Army staff member's name appears on both the official bidders meeting attendees list and the list of participants at the Baghdad planning workshop, which was included in the document. Nevertheless, careful study of the official transcript of the bidders meeting in Dallas demonstrates that the Army staff present knew about the plan as well.[15]

The plan was officially approved on July 24 by General Crear, Phillip Carroll, the Coalition Provisional Authority's Senior Oil Advisor, and Thamir Abbas Ghadhban, CEO of the Iraq Ministry of Oil. What was posted on August 1 strongly suggests that the Corps's Fort Worth staff had received the plan before July 24 and could have passed it along to bidders long before they did. The 33-page document's approval page was unsigned, but a separate document posted simultaneously consisted of the approval page with the Crear, Carroll, and Ghadhban signatures and the July 24 date handwritten. If the plan and the signed approval page had not been received on different dates by Corps personnel in Fort Worth, they would not have been separate documents. The plan had probably been finished well before July 24, in order for it to be circulated for review and approval by the Corps, the Coalition Provisional Authority, and the Iraq Ministry of

Oil. The Corps could have posted the plan for bidders to see as soon as they received it, but the Corps delayed the posting until August 1, less than two weeks before the bidders would need to ship their proposals to the Corps's District office in Fort Worth.

Why was the *Restoration of Iraqi Oil Infrastructure Final Work Plan* hidden from bidders as long as possible? Because it proved that the full and open competition was a fraud. The document and an accompanying ten-page spreadsheet showed that the Defense Department had already committed the bulk of the remaining Iraq oil work to Halliburton KBR. The workshop participants had identified 220 projects that needed to be completed in order to restore the Iraqi oil production infrastructure to its pre-war capacity. The cost estimates for the 220 projects totaled $1.144 billion, and Halliburton KBR's share of the work would total $950.3 million or 82%. The remaining 18% or $193.3 million worth of work would go to the Iraq Ministry of Oil to administer directly, mostly through its engineering and construction subsidiary, the State Company for Oil Projects. Halliburton KBR's 82% share included projects it would manage directly, as well as equipment and materials it would purchase for projects executed by the Ministry. It also included the costs of technical assistance, management support, and other services it would provide to the Ministry. Of Halliburton KBR's $950.3 million share of the total work, approximately two-thirds was for commodities, which meant easy money. When $630.4 million worth of purchase orders for equipment and other items go through the company's accounting system, a percentage of the costs is added as the company's fee. The total budget was actually $1.485 billion because $193.5 million for contingency was added to the $1.144 billion total cost for the 220 projects.

According to the schedule outlined in the *Restoration of Iraqi Oil Infrastructure Final Work Plan*, of the $1.144 billion for projects, $716.2 million worth of work was scheduled for completion by the end of September 2003, another $250.9 million worth by the end of December 2003, and the remaining $176.4 million worth by the end of March 2004. Even if the Army stayed on schedule with the competitive procurement and announced the two winners on October 15, by the time the contracts were actually negotiated and signed and the contractors mobilized their personnel to Iraq, it would be January. By then, $967.1 million worth of the total work planned would have already been completed by Halliburton KBR and the Iraq Ministry of Oil, or if not completed, at least committed and not available to the new contractors. If the winners were not announced on October 15, there would be even less.

Before returning to the bidders meeting in Dallas on July 14, statements in the *Restoration of Iraqi Oil Infrastructure Final Work Plan* confirm beyond any doubt that the Defense Department intended for Halliburton KBR to continue as the Army's sole RIO contractor. In the project execution section of the document, the paragraph headed "Kellogg Brown & Root (KBR) Commitment" states in no uncertain terms that KBR is the Government Operations Contractor selected by the Army for the RIO program and that "support of the USACE [U.S. Army Corps of Engineers] commitments to the Iraq Ministry of Oil can be made via the KBR RIO government contract." The paragraph headed "Procurement — Definition & Execution" states, "It will be most prudent for KBR to manage their primary planning and initial procurement work from Kuwait City due to communications, travel, and security arrangements currently in place. Later as freedom of movement and telecommunications can support transitioning of procurement activities to a more central control point in Baghdad..." This statement suggests that the plan was written long before July 24 because by then Halliburton KBR, Bechtel, and other U.S. Government contractors had been working in Baghdad for several months.

Finally, in the program management plan section of the document, the paragraph headed "Purpose" states, "The restoration of the Iraqi oil infrastructure will be executed as a team effort. The team will include the Iraq Ministry of Oil, the Ministry's subsidiary operating and engineering construction companies, OCPA [Office of the Coalition Provisional Authority], USACE-RIO, and the U.S. Government Services Contractor, KBR." The organization charts included in the document confirmed that Halliburton KBR would be involved in every aspect of RIO project execution in both the north and the south of the country and even on the board that would measure the success of the project execution team of which it was such an integral part!

Meanwhile, back at the July 14 meeting in Dallas, the bidders were restless...

All except Stoney Cox and the other Halliburton KBR personnel, who were heard to laugh or snicker at various points in the meeting. They knew it was a farce. It is evident from the meeting transcript that the Corps personnel who conducted the meeting or answered questions from the floor also knew the results of the joint planning meeting held in Baghdad the previous week. They had to know that nearly $1 billion worth of work was already committed to Halliburton KBR and would be completed or well on its way to being completed before the two new contractors could claim their shares. The official transcript of the meeting, also known in government procurement as a pre-proposal conference, showed that that the Corps's legal counsel was nervous.

Robert Slockbower, the Military and Technical Director for the Corps's Southwest Division, opened the meeting as emcee and introduced Colonel Schultz.

The acting commander began, "On behalf of General Robert Crear, the commander of Southwestern Division and currently commander of Task Force RIO, I want to welcome you here to Dallas and our pre-proposal conference here this morning. I would like to say that General Crear probably would have preferred to be here today and welcome you personally, but I think he's having too much fun currently in Iraq, with meeting the challenges there and working with the great civilians, soldiers and contractors that we current have on board there, working to restore the Iraqi oil system for the people of Iraq... Today the mission continues to focus on importing fuel products, restoration of the oil system itself, involving all of the myriad of pipelines and systems and other equipment associated with that effort over there, restoring communication systems to help with the export of the product, and replacing looted equipment, which has been a significant challenge for us over there..."

"Task Force RIO was established in Kuwait just ahead of the start of hostilities. Our initial effort concentrated on getting a contracting capability or a contractor capability for firefighting set up and operational immediately ahead of the hostilities... We were fortunate that the damages were not as severe as anticipated, and you've seen a lot of those projections in the newspapers in how the press covered that. Many of the facilities, we were fortunate, were captured intact. We extinguished nine oil well fires with the assistance of the Kuwaiti Oil Company, who volunteered early on to help with that effort. As hostilities diminished, we began making emergency repairs, and concentrated on minimizing the damages to the environment, which were already occurring over there. As the security and safety conditions allowed, we began making damage assessments and initiating repairs to get the overall system operational and first oil flowing in both the North and the South."

He talked a few more minutes and then said, "We're about answering your questions in this pre-proposal conference. I'd now like to ask Gordon Sumner, our Director of Contracting for Southwestern Division, to come forward and talk you through some of the acquisition strategy and a little bit about the process ahead."

"Like any plan," Sumner began, "you have to put together your milestone chart which starts with your strategy. Believe it or not, we've come a long way already."

At that point, no one in the audience, except the Halliburton KBR contingent, would have appreciated Sumner's remark.

He continued, "We have a long way to go, but today is another milestone that we're crossing that we're very proud of. Our acquisition strategy was approved in June, and the plan, which was more specific to exactly what the Corps of Engineers Southwestern Division will be doing, was approved roughly a week after that."

None of the bidders, except Halliburton KBR, could have known for sure that the acquisition (aka procurement) strategy to which he referred was the document that Halliburton KBR had prepared, the document presented as the scope of work in Section J, later moved to Section C, in the RFP. At this point in the proceedings, the bidders assumed Sumner simply meant the competitive acquisition of two new contractors, one for the north and one for the south. None of the bidders, except Halliburton KBR, could have known about the "unified strategy" that the Army, Halliburton KBR, the Iraq Ministry of Oil, and the Coalition Provisional Authority had developed at the joint workshop in Baghdad that ended just 5 days before. At this point in the Corps's procurement process, after all the outcry over political cronyism and Halliburton KBR's sole-source contract, the bidders assumed the Army was finally playing it straight and they had a fair shot at over $1 billion worth of work.

Later, Sumner addressed the question, "When is the award expected?"

"I can't push this enough," he said. "We are on an aggressive schedule. Some of you that are used to doing government contracting and making proposals, know that some of those processes can last up to a year. We're hoping to have an award by mid-October, with people on the ground and in country at the end of November and early December turning wrenches."

The final question Sumner addressed was, "How does base fee and award fee work, and how much can we ask for?"

"A fee is just that," he said, "it's a fixed amount of dollars that's negotiated between the contractor and the government during discussions prior to award. This amount is currently — and we plan to negotiate the fee on every separate delivery order." Sumner had almost violated the most sacred trust that a contractor has in the government — not to divulge the company's base fee percentage to competitors, since that is what often determines the winner on cost-reimbursable contracts. Halliburton KBR would have been incensed.

"Again," he said, "we urge you to propose what you feel is reasonable, and it is a negotiable item if we get to negotiations, when we get to negotiations rather." It was another slip.

Soon Robert Slockbower became emcee again and asked the members of the panel assembled to field questions from the floor to introduce themselves. The panel consisted of Director of Contracting Gordon Sumner, whom the bidders had already met, Defense Contract Audit Agency (DCAA) Regional Audit Manager Tim Carr, Chief of Technical Engineering and Construction Darrell Alverson, and Southwestern Division Counsel Morris Tanner.

After introducing himself, Tanner said, "I have a couple of comments I'd like to make before we get into the general questions. The first is that we have a lot of uncertainties in this contract. We have put out on the Web for you all to look at a Rough Order of Magnitude [ROM cost estimate] which has been developed by our Team RIO over in Iraq, and that does not represent the work which we anticipate will be done under the contract you're competing for. It represents the sum total of the work that we know is out there as of the time the Rough Order of Magnitude was completed, which is about three weeks, three or four weeks ago."

Actually, the Rough Order of Magnitude estimate attached to the RFP was dated June 3, six weeks before. So the essential division of the work between Halliburton KBR and the Iraq Ministry of Oil had not just been decided the preceding week at the workshop in Baghdad. If the estimate did not represent the work to be done under the new contracts, why was it included in the RFP? Especially when the RFP said, "The accompanying ROM covers the work required to execute this mission. Damage assessments provided by the current sole-source contractor formed the baseline from which government personnel prepared the ROM... The ROM shows the amount of work projected to be done for the repair and restoration of the Iraqi oil infrastructure, its rough dollar value, and the schedule for accomplishing that work."

The Army's statements to the press and letters to Representative Waxman had been full of contradictions for months. There were so many contradictions between the various documents posted for bidders and between the documents and the dialogue at the meeting that bidders did not know what to do. It was not clear what scope of work they should address in their management and technical proposals and what scope of work they should use as the basis for their staffing plans, cost estimates, and bids. The RFP went on to say, "The schedule in the ROM represents the government's best estimate for the completion of all necessary repair and restoration. Under this schedule, a significant amount of the work needed is likely to have been begun under the current sole source contract before a competitive contract can be awarded." But the bidders in Dallas still believed at this point in the procurement that the work would be transferred from Halliburton KBR's sole-source contract to the two new contracts once they were in place. Director of Contracting Gordon Sumner had repeatedly stressed that the Corps was on an aggressive schedule to do just that.

Tanner continued talking about the amount of work and the schedule presented in the Rough Order of Magnitude document included in the RFP, "That can change. Some of that work, of course, it contains a schedule, and a lot of that work will have been awarded by October, even if we're successful in getting this contract awarded in October."

Of course, any experienced government contractor in the room knew that but still had no way of knowing yet that it was a done deal — except the thirteen representatives of Halliburton KBR. Perhaps this was one of the many moments in the meeting when they were heard laughing quietly among themselves.

Perhaps realizing he had gone too far, Tanner then said, "There are things that can change the amount of work that's available. As you know, there continues to be sabotage over there. There may be intentional sabotage. There may be more looting. Some of that could happen. Damage to a large plant could affect drastically the amount of work which is available."

Tanner should have stopped there, but he did not. He joined the other Army staff in the charade, "What we have done is provide you the best information we can on the type of work which is available and the best information we can on when we think that work will be awarded as of three weeks ago."

"Having said that," he continued, equivocating again, "I will also say that another factor that can affect this is the capabilities of the Iraqi Oil Ministry and the companies that work for it. There are a number of political decisions being made, some of which encourage very strongly the use of these companies to do whatever they are capable of doing. So the ability they have to reconstitute their work force and do work may end up affecting the scope of our contract. We do not know at this point exactly how quickly they will reconstitute their work force and be able to execute some of the work that is done here."

Those political decisions had already been made at the workshop in Baghdad, if not before, and according to the press, the Iraqis working for the Ministry of oil had been doing a lot of emergency repairs and fuel delivery themselves. The Final Work Plan confirms the Defense Department's belief that the Iraq Ministry of Oil was capable of the work it would undertake on its own but with U.S. taxpayer funds.

Tanner pressed on, equivocating once more, "another thing that may affect us is funding. You will no doubt notice that we have issued this solicitation subject to the availability of funds. There is a dollar figure in the Rough Order of Magnitude estimate of $1.68 million. That includes 20 percent contingencies and could vary up or down by a 40-percent factor. As of this time, we have been provided a total of $489 million by the Congress as appropriations."

But all the bidders knew Bush was requesting billions more.

"If we don't have enough money to do the work which is necessary to be done by October we obviously won't be letting as many of the contracts as we would otherwise. Those are the only comments which I think I need to make at this point," and Tanner sat down.

At that point, emcee Slockbower asked for someone "to break the ice with question," and a series of bidders eventually stood up, but the most pugnacious were representatives of large savvy companies like WGI and Fluor.

WGI's John Dougherty was the first to ask about the two sample tasks in the RFP, one involving firefighting and the other damage assessments, two of the three activities that everyone knew were already being performed in Iraq by Halliburton KBR (the third was emergency repairs). As the current contractor, the company would have a clear advantage in the evaluation over everyone else. "What is the relation of those two pro forma tasks to the actual? Are they anticipated to be embedded into the contract or are they to develop a test case, so to speak, for evaluation purposes?"

Darrell Alverson, Chief of Technical Engineering and Construction, responded, "It's actually put in there as a sample cost proposal. The costs, obviously, are not binding because it's a cost-plus contract and the actual task order will, in fact, dictate what you provide. But those are, in fact, two actual things that we will ask for, and are also contained in Section J of the proposal or the scope of work itself. But those will be real provided services, yes, sir."

Alverson's was a strange response, given that all nine well fires had already been put out, but stranger still because he openly admitted that the sample tasks were meaningless exercises. It was meaningless to the Army if the results were already known, but the cost estimates would take other bidders hundreds of hours to complete.

Next up was Ken Oscar from Fluor, who turned out to be the most perceptive and persistent questioner of all. He began by pointing out, "Throughout the RFP it refers to a nine-page Statement of Work in Section C. There is no Section C. There are several things in Section J, one of which is an eight-page, looks like a — I don't know what it is exactly."

"A menu," Gordon Sumner, Director of Contracting, responded.

"It looks like a statement from a previous contract you had," Oscar continued, "As a matter of fact, the top of it says, 'LOCAP.' Then there's a ROM [Rough Order of Magnitude]. What is the Statement of Work?"

Gordon Sumner responded again, "It's actually that listing in J. And again, that listing is a menu. We don't know today what requirements we will have to draw upon, which task orders within that menu that we will need, but you need to be prepared to do everything within that eight-page, nine-page menu. And the wrong..."

"So the first document in J is the Statement of Work?" Oscar interrupted.

"Yes. You are right. Probably for ease of finding it should have been right there in Section J. It's only nine pages but it is part of the Section J Statement of Work, nine pages. That is the Statement of Work." This response suggests that Sumner was getting flustered, for every government Director of Contracting knows the Statement of Work is always in Section C — that's Proposal 101 — and this Director of Contracting should have known his own RFP.

After three other questions had come and gone, Sandy Davis from Fluor stood at one of the microphones with a series of his own. One was "In Section L you ask us to provide you with up to ten project examples of work performed, past performance and experience. In Section M you indicate you will evaluate five of those. So I assume that's a typographical error or are we going to submit ten and you're going to evaluate five?"

John Rodgers, in the room but not a member of the panel, answered Davis, "We'll have to look at that. Whatever you submit is what we will evaluate."

As Contracting Officer for the two new contracts, Rodgers should have known the RFP better than anyone, but he did not. Clearly, some of the bidders had analyzed it with a fine-tooth comb.

Davis continued, "Okay. Page count question, if I may. Again, [I] understand the need to limit the proposal to a hundred pages, but you have asked us to limit each of the project descriptions to two pages. So ten projects times two, that's twenty pages already. And if we put in as many as eight or ten resumes, 30 percent of the page count is gone just in projects and resumes. I'd ask if we can consider exempting either the resume page count and/or the project page count from the hundred so we can better explain our management and technical approach."

At that point, Davis paused because the panelists began a discussion among themselves, not audible in the rest of the room. When he continued, his questions were addressed by Gordon Sumner, but his answers were the same as Rodgers' — "We'll get back to you on that."

Davis was not dissuaded from continuing to prod. "Okay. A little testier. On the task-order-driven contract, particularly internationally in an environment in Baghdad, can we anticipate that you will or will not fund a program management office to be established in country to support the contract?"

All the Corps personnel were jumping in now, and Alverson answered Davis with a rather cryptic reply, "Whatever is needed to do business."

"Yes, whatever your technical solution is," Gordon Sumner added.

"Okay. Again, it's an ID/IQ [indefinite delivery/indefinite quantity] contract," Davis reminded them, "so if we feel a need for a program management in country then we need to propose that and assume it will be funded."

A brief inaudible utterance was the only response. The Corps was clearly unprepared, but why, given the procurement's high visibility? One can only conclude that they were not taking it seriously because they knew it was a sham. Halliburton KBR had all the work, and no new contract would change that to any significant degree, if at all.

The meeting continued with the questioners about evenly divided between those attempting to understand where they might fit in, such as one Fadi Nammour representing a large Iraqi company, and those, such as Fluor's Ken Oscar, still trying to figure out what the Army expected in response.

"Section L, Paragraph 7, Subcontracting Plan," Oscar began. "The section is labeled 'A Subcontracting Plan' as Volume 3. Prior to that, there's a different description of what should be in Volume 3. It appears like there's two Volume 3s. Will the Subcontracting Plan be a separate volume, not in the hundred-page count, or is that supposed to be embedded somewhere?"

"You're going to get a personal answer right here," Darrell Alverson, Chief of Technical Engineering and Construction, laughingly replied.

"Yes, I just looked at the RFP," Contracting Officer Rodgers said, as if for the first time. "The Subcontracting Plan, Volume 3, is misstated. There is no separate volume for Subcontracting Plan, so it will just be part of a...I'm going to say right now; we will change it by amendment. It will probably be an addendum that won't count for the hundred pages." An answer that began as if it would be definitive ended with another equivocal deferral.

Oscar tried again, "Kind of a follow-up to that. The original advertisement for solicitation said that it was very important to employ Iraqis and Iraqi companies. There is no mention of that in the RFP, and I was wondering... what the RFP does stress is U.S. subcontracts, so could you explain a little bit between hiring Iraqi companies and hiring U.S. small business companies?"

"The Federal Acquisition Regulation has no allowance for giving extra weight to a proposal for, for instance, an Iraqi company," Gordon Sumner, Director of Contracting and Rodgers' boss, began. "It does allow for subcontracting to small and small disadvantaged business firms. There may not be any small and small disadvantaged businesses that are in the oil services field that you intend to put in your part of your technical solution, and so that's why we can't say what the minimum amount of participation by small and small disadvantaged businesses [will be]. Again, it's not a requirement over there, but it's something that we will look at. But, on employing Iraqi companies, it's encouraged and that's what our direction has been, but there's no provision in the FAR that we can weigh that as part of the technical solution evaluation. It is complex."

No, it appeared rather simple. Bidders would be scored on something that was probably not doable and not scored on something that was admittedly important. Still, most bidders remained optimistic. Only a few were starting to wonder now whether the whole procurement was simply a farce.

John Allen, representing the International Response Corporation, continued the theme, "It's been the notion, at least from the synopsis you presented, that we should be striving to identify host country nationals, and yet I hear on one hand that that's not a weighted factor, and yet it's still encouraged. I really need to know where, because a lot of effort is going into doing those sorts of things. Could you please clarify a little bit more? What is the value of efforts of companies to use host country nationals versus third country nationals, or U.S. citizens?"

Legal Counsel Morris Tanner replied, "Well, again, this has been a matter which has been discussed a number of times at the policy level in DoD [Department of Defense] and in Army, and those are the instructions that resulted from those discussions. Now, we do not have authority in the FAR to give extra credit, if you will, for companies from various — from any country that we happen to like today, so we don't have any authority to do that. On the other hand, we do have political goals in the United States to encourage the participation of Iraqi companies and Iraqi workers, so that's what you see in the solicitation. It is somewhat ambiguous, but that's the way it is."

And ambiguous is the way it remained, with the bidders in a quandary as to how to do the right thing. And a few minutes later, the questions that all of them had wanted to ask was finally posed by Juan Llovet from Parsons, another one of the big five.

"How will the work be transitioned between the incumbent contractor and the successful bidders? How do you visualize that transition?"

Darrell Alverson, Chief of Technical Engineering and Construction, fielded the question with another of his rather cryptic replies, "Kind of hard to visualize because it's several months out, but obviously, when we would have to assess each item of work on a case-by-case basis as to whether it's advantageous to leave the current contractor in place to finish up. I can't give you a good, sincere answer other than to say we would evaluate each area of work on a case-by-case basis and then transition the new work to the new contracts and start there."

And he was right. He could not give a good, sincere answer because the bidders would have gotten up and left.

Llovet continued, "If I may be allowed to follow on then, does that mean then the existing incumbent contract will be in full force at the time of award of these two contracts?"

Gordon Sumner answered before Alverson could, "At the time of award, correct. We have to keep someone on the ground supporting the restoring of the oil infrastructure. It would be a transition, but yes, they will be there until we have new contractors on the ground turning wrenches. It's just not business-smart to pull them out any sooner."

Or perhaps not smart for one's career. So there it was, the handwriting on the wall, as clear as could be. Yet the optimistic bidders still did not fully recognize it for what it was, or perhaps they were willing to grapple for the bones.

Fluor's Ken Oscar was still grappling with the RFP, beginning to grasp the meaning of what he had heard, and would not let up. "I'd like to go back to the Statement of Work, the ROM and the Section L. There seems to be differences between the three. Section L, Paragraph 6.2.2, Subfactor 2, for example, says 'cradle to grave.' The ROM excludes from the wellhead down. The Statement of Work, which is labeled 'Concept,' and on the first part of it says, it appears to be the original statement that was given to the LOGCAP contract, excludes some different things. So what exactly is the precedence? Is it the first thing that's an eight-page Statement of Work, or is the ROM just meant to be information, or is it Section L?"

"I think the best we can do, review and clarify again, just make sure we state what the relationship is." Of course, the ambiguous response could only be Alverson's again.

Amazingly, Morris Tanner, the Corps's Legal Counsel, knew what he meant. "Right," he said. "The ROM, again, is not necessarily intended to be the work, which will be done under this contract. It has been developed by our Team RIO over there in order to scope out the work that is now required, or was required a month ago when this thing was finished. A lot of that work will probably have been done by the time this contract is awarded. There may be other work which comes up because of additional sabotage, looting or whatever. So, the scope of work in the contract, anything which might happen in the future, which we don't currently see over there. The ROM indicates our best estimate of what we currently see over there."

Fluor and the other bidders needed to know exactly what scope of work to address in the technical and management approaches they were required to describe in the proposals due August 14, so Oscar attempted to summarize what he had just heard, "So the eight-page Statement of Work, it's pretty restrictive, but that was meant to be broad. The ROM is meant to be your estimate as you saw it a couple months ago. And then Section L, that's the broadest of all, that really says 'cradle to grave.'"

"That's correct," and that's all Tanner said.

Fluor's Sandy Davis tried again but could not elicit any more definitive response. A few minutes later, Davis changed the subject and asked, "Are contracts being awarded directly from the Iraqi Oil Ministry for the northern and southern oil fields, and if so, where are those funds coming from?"

"At the moment," Tanner began, "we're not aware that any such contracts are being awarded by the Iraqi Oil Ministry. The exports, which are now starting, will produce cash, which will be available for the Iraqi government, and I would assume that at some point in the future, the Iraqi Oil Ministry will begin awarding some of its own contracts. When that will happen, I don't know."

Of all the Army personnel at the meeting, Tanner had been the one who came closest to presenting the truth — agreed on at the joint workshop in Baghdad the week before — until now. The Defense Department had agreed to make $193.3 million available for the Iraqi government, through the Ministry of Oil, to award its own contracts for some of the RIO work.

Later in the morning after a break, someone asked, "In the scope statement, the first item is "Elements Under LOGCAP" and it has to do with prepositioning equipment and people, was that intended to be done under LOGCAP or scope items that are included under this contract?"

"The direct answer is both." It was Alverson responding, of course. "We provided those things in a pre-position mode. Now some of those things are existing over there. But the capability is desired under this contract."

The questions and answers became shorter and more direct, and before the meeting was over, Ken Oscar asked, "Can a single contractor win both contracts?"

Morris Tanner started to answer, paused, and then said, "The solicitation says 'two contracts.' It doesn't say 'two contractors.'"

"Okay," Oscar laughed.

"I got a quarter. You want to flip it now or later?" Tanner quipped. In retrospect, it may have been the most honest response from the Corps all day.

Emcee Robert Slockbower quickly put an end to that, "A question from the back, sir?"

A long line of questioners asked for clarifications, pointing out more and more contradictions and discrepancies in the RFP. Finally, someone asked, "I guess it's Section J where we begin to introduce the scope of work for the contract — and I'll quote from it — 'Corps of Engineers IDIQ Contracts. Two contract procurements with identical Statements of Work, one sole source and one or more awarded through limited competition.'"

"It's a misprint," Gordon Sumner, Director of Contracting, said. "It's going to be two fully competitive awards made. Like Morris said, two contracts awarded, let me just put it like that, full and open competition."

The questioner continued, "So, essentially two contractors will be expected to provide the full scope of these 13 items, more or less?"

"Two contracts, contractors, yes." Sumner had slipped again.

The questioners turned to other matters for awhile until Charles Sweeney from Foster Wheeler asked, "Do you see that there will be any other initiatives going on in Iraq, either through yourselves or through other agencies or other government groups, multinational groups that may overlap or interfere with the work scope which you're outlining in this inquiry?"

"I can't say for sure," Sumner began, "but I can tell you one of the reasons why we've decided to do two contracts, one North and one South, is so that we won't have two contractors stumbling over each other." By that logic, they should only award one new contract, or even better, continue the one they had.

Just before the meeting ended, Fluor's Sandy Davis came to the mike. "Just for the record, the last question. One of our worthy competitors you already have under contract over there and have had for a number of months. It happens that they've already performed all of the tasks that you ask for in the sample projects and can pull that information obviously off the shelf. So the direct question is, has the government performed a conflict-of-interest determination and convinced yourself that there's no competitive advantage to the incumbent?"

Counsel Tanner answered. "The requirements in the new contract," speaking in the singular again, "were developed by Team RIO. The ROM was developed by Team RIO. We have looked extensively at the conflict of interest or organizational conflict-of-interest issue. So has Army and DoD. Everyone is convinced at this point that there is no organizational conflict of interest." Tanner did not say why.

Everyone on the stage may have been convinced for one reason or another, but no bidder in the audience was, and when Team RIO's Restoration of Iraqi Oil Infrastructure Final Work Plan was finally posted on the Web on August 1, there was the proof. Halliburton KBR was deeply embedded with General Crear's Task Force RIO in Iraq.

6

Questions & Answers: More Handwriting on the Wall

AFTER THE BIDDERS MEETING, the government posted a series of questions and answers (Q&A) on the Web.[15] The questions had been e-mailed to the Corps by individual bidders, but the answers were made available to everyone. The first posting on July 21 was called "Questions for Clarification," but the answers it contained were anything but clear; some confirmed what had been said at the meeting, and some contradicted it. But the message was becoming clearer when seen in black and white. For example, the answer to one question said, "It is not possible to predict accurately how much of the work described in the ROM will have been completed by the time the follow-on contracts have been awarded and the contractors mobilized in Iraq." The answer to another question said, "Political decisions which may be made between now and the time the new contracts are in place may affect whether a task order is issued under those contracts to extinguish the fires." The Corps was referring to any new fires started by sabotage, but "political decisions" suggested they would apply to other work as well.

It was the Army's turn to become testy now. A very valid question was asked: "The ROM states that, 'Oil ministry will supply the majority of the labor for the repairs and operations.' Please define what is meant by majority with respect to all of the task orders to be issued after award. Does this eliminate the need for self-performance and subcontractors?" The Corps's answer was nonresponsive: "Majority is defined in the dictionary. Once again, the mission assigned to SWD [Southwest Division, which includes the Fort Worth District] essentially was to repair and restore the Iraqi Oil Infrastructure to pre-war conditions. SWD has no authority to engage in operation of the Iraqi Oil infrastructure. To the extent necessary, technical assistance will be provided to the Iraqi Oil Companies, who have the mission of actually operating their facilities." The questioner was clearly concerned with pricing labor for the sample tasks, but the Corps did not address the labor for the repairs.

On July 22, more Q&A appeared on the Web. The first question was, "Given their obvious competitive advantage and current positioning with the Corps and in the region, has a formal OCI [organizational conflict of interest] determination been made regarding KBR's ability to bid this follow-on contract, if so can we be provided a copy? Will KBR be allowed to bid?" The answer was anything but clear, despite the caps; in fact, it was outright obtuse: "THE FORMAL OCI IS CONTAINED IN THE ACQUISITION STRATEGY WHICH IS PROTECTED FROM RELEASE BECAUSE IT IS PRE-DECISIONAL."

In a second posting on July 22, Question 3 was actually a request: "The initial section of this SOW introduces an 'impossible specification' as a contract requirement. Since KBR is the holder of the LOGCAP contract and since the delivery dates have passed, it is impossible for other contractors to respond to these requirements. Recommend that this first section of the SOW [statement of work] be deleted as a requirement for this solicitation." The Corps answered, "Amendment 0001 modifies the Work Statement (Section C) to remove the language described in Question 3."

Question 4 also asked about the SOW, specifically Task 7, Manage Oil Reservoirs, and Task 8, Internal Distribution and Sales of Refined Product: "Inclusion of these services in this SOW appears to contradict the intent of this solicitation for several reasons: reservoir management is not a short-term restoration activity; Iraq oil reserves are an Iraq national asset; national reserves data is typically proprietary information; these services appear beyond stated U.S. objectives. Recommend that these two items be excluded/deleted from the SOW for these contracts." In answering, the Army's disdain for the bidders was the only thing made crystal clear: "As stated in the pre-proposal conference, the Government seeks a capability to provide services such as described in Question 4. Requiring a capability and using the capability as a technical resource, subject matter expert or to provide technical assistance are all requirements under this contract, whether or not a task order ever asks for these services or not. The requirement is valid as stated." And the media, corporate watchdogs, and Representative Waxman complain that contractors waste taxpayers' money. In this case, it was the other way around. The Army would make contractors jump through a series of needless hoops, their proposal budgets be damned.

By the next day, all civility on the part of the Army Corps of Engineers ceased. Question 17 on July 23 expressed a legitimate concern: "Will the Government provide a summary of the latest acquisition policy changes that may affect awarding or performing work in Iraq? The Commerce Department, for example, has reported that the 'Buy American Act' has been waived for contracts issued for support in Iraq." The Corps's answer was again in caps: "READ THE SOLICITATION AND LAW, MAKE YOUR OWN ASSESSMENT." A simple yes or no, as to whether the Act applied, would have sufficed.

Given the continuing attacks against American personnel after the conflict ended, Question 19 had made the following request: "The RFP provides Force Protection to Contractor employees through the Service Theater Commander, but does not give the Contractor the opportunity to act on behalf of its employees' best interests when conditions are considered to be unsafe. In view of the continuing unstable conditions in Iraq, we request that the following clause, which has been included in agreements for other ongoing work in Iraq, be added to the solicitation as a new H [contract] clause: 'Contractor shall have the right to notify the Contracting Officer and withdraw its personnel or postpone a job order if it is determined that current conditions will be unsafe from a security or safety standpoint due to instability in

Iraq. Such withdrawal, postponement, or declination to perform a particular job order will be deemed to be an excusable delay in accordance with the FAR clause at 52.249-14, Excusable Delays." And the Corps's reply? "There is no need for this; the Government will be reasonable in its approach to contractors." The Army had been anything but reasonable; it had been arrogant, and the pattern continued.

Question 20 made another legitimate request: "Clause H.25, Method of Ordering (Unilateral and Bilateral), gives the Contracting Officer the right to unilaterally issue a task order. While it is understandable that a situation could arise such that an unplanned task order is needed, there are no stated constraints on when or how this approach would be used. Potentially this puts the Contractor at risk of having to perform unanticipated and high-risk work, perhaps even at a fixed price, without negotiation. To avoid this imbalance we request that the unilateral option be deleted." The Corps's answer was simple: "The unilateral option will remain."

The July 23 Q&A ended with the following telling exchange. Question 24: "Could you please clarify what is meant by Section 5 in the Scope of Work, 'Provide Logistics Support to the Government'?" The Corps's response: "PROVIDE LOGISTIC SUPPORT TO THE GOVERNMENT IS SELF EXPLANATORY." And to the Army's LOGCAP contractor it was. If the Defense Department had any intention of bringing a new contractor on board to supply General Crear and his Task Force RIO with logistics support in Iraq, as Halliburton KBR had been doing for months, you can be sure the Army would have answered Question 24 in a different way.

The next Q&A was posted on August 6, after the *Restoration of Iraqi Oil Infrastructure Final Work Plan* had appeared. Question 1 came right to the point: "This Final Work Plan was developed from a 6-9 July Workshop in Baghdad yet no mention was made of this Workshop at the 14 July CESWD [Corps of Engineers Southwestern Division] Pre-proposal Conference. Several KBR representatives who participated in the Baghdad Workshop also attended the pre-proposal conference. Why was this Workshop not made public at the pre-proposal conference?" The Corps answered in three sentences, and all three were, frankly, lies: "There are frequent meetings of this type involving RIO personnel, Iraq Ministry of Oil personnel, Coalition Provisional Authority personnel, and, on occasion, contractor personnel. It was far from certain when agreement would be reached on a final work plan. The plan itself was made public as soon as reasonably possible after it was finalized." Perhaps the third sentence was true; withholding the plan until 2 weeks before proposals were due was "as soon as reasonably possible" if the Army wanted no new contractor to have a chance.

Question 2: "The requirements of Section C and the transcript of the pre-proposal conference requested world class contractors/teams capable of providing cradle to grave capabilities on work up to $500 million over 5 years. If the role of CESWD [Corps of Engineers Southwestern Division] contractors is now to simply augment or compliment the work being performed by SCOP [State Company for Oil Projects, a subsidiary of the Iraq Ministry of Oil] and KBR, our level of participation will be dramatically reduced and our management approach will be much different. Please clarify the expected role of the CESWD contractors in the North and South of Iraq with regard to the role of SCOP and KBR." The Corps replied: "We are unable to provide any clarification at this time with the exception of the final work plan which has been provided. As explained at the pre-proposal conference, there are a number of unknowns which will influence how much work may be available under the competitive contracts." Perhaps those unknowns included how long the Defense Department could prolong the procurement so there would be no work or funds left uncommitted by the time a new contractor came on board.

Question 3 tried again: "The requirements of Section C and the need for world class cradle to grave capabilities appear to be inconsistent with the role and responsibilities established in the Final Work Plan. The Corps's answer was, "The Final Work Plan is not the statement of work for the services to be performed under the two new contracts." If the Final Work Plan was the final plan for the work to be performed, then clearly there would be no work performed under the two new contracts.

By the time the last Q&A was posted on August 6, Bechtel had already dropped out; realizing the procurement was a fraud, they were cutting their losses. Parsons and Fluor stayed in, but the number of other proposals the Corps received the following week has not been released. The companies that did bid did so at considerable expense. Preparing a government proposal is a major effort, and some companies spend millions of dollars to bid on large federal contracts. All indications were that there would be little return on these, so why did Parsons and Fluor persist?

"If you look at the oil industry in Iraq and its 20 to 25 years of neglect, there are certainly billions of dollars of opportunity in getting it up to speed — and then expanding it," Ron Oakley, Fluor's government contract business head, said at the time. "Anytime our folks expend as much time and effort and energy into responding to a [request for] proposal as they have here, they go in it with the expectation of winning, anything less than that is a disappointment."[16]

On August 13, the day before the proposals were due, an Army Corps of Engineers spokesman, Bob Faletti, told the *Los Angeles Times* that "there might not be enough emergency repair work to merit additional contracts."[16] He said "the main reason the Corps was accepting bids for two new oil contracts was the possibility of more looting and sabotage." So all of the original work identified, more than a billion dollars' worth, had indeed been divided between the Iraq Ministry of Oil and Halliburton KBR. The Army had used the promise of that work to bait the bidders and then switched the prize the day before the bidders' proposals were due. The new prize was worth nothing, unless the looting and sabotage continued, and maybe not even then.

7

Contractor and Corps Improprieties: The Finger Pointing Begins

AT A BRIEFING IN BAGHDAD on August 25, the Coalition Provisional Authority confirmed what some of the bidders suspected: Halliburton KBR could be awarded most of the remaining work before new contracts were awarded in mid October.[17] So the CPA had gotten the word from the Army (or vice versa). By early September, the Army had issued task orders to Halliburton KBR under its sole-source Iraq oil contract that totaled $948 million, and President Bush had requested an additional $2.1 billion for Iraq oil field reconstruction work. Representative Waxman requested a briefing from the Army Corps of Engineers to explain the need for the additional funds, but the briefing was not held. The Corps told Representative Waxman that they had not been involved in preparing the new budget request![17]

October 15 came and went with no announcement by the Army other than they would need two more weeks. On October 29, the Army issued an amendment to the RFP asking bidders to submit proposal revisions by November 5.[15] It said, "This amendment is necessary to increase the contract capacity, but will not change the scope of work. As Task Force Restore Iraqi Oil (TF RIO) has completed repairs, it has conducted increasingly thorough analyses of the situation on the ground, to include the effects of looting and ongoing sabotage. The Task Force has also been able to analyze the independent needs of the southern and northern oil fields and supporting infrastructure. As a result, TF RIO has identified more oil infrastructure work requirements than originally envisioned."

As luck would have it, while Task Force RIO was finding more work, Congress was authorizing $18.6 billion in additional funding to rebuild Iraq.

The amendment continued, "It appears that import and distribution of refined products currently being accomplished under the sole source contract, will not be accomplished under these replacement contracts." Halliburton KBR's overcharging for fuel had been in the news for months. They had been charging the Defense Department an average of $2.65 per gallon, while another supplier of gasoline to the U.S. military in Iraq, the Defense Energy Support Center, had been charging no more than $1.19, and the Iraqis were doing it themselves for 90 to 96 cents; all three prices included transportation costs.

Nevertheless, the Corps's amendment said "the scope of work remains broad enough to include this requirement to deal with contingencies... The estimated maximum amount of the subsequent contract is increased from $500 million each to $1,200,000,000 for Southern Iraq and $800,000,000 for Northern Iraq."

So the ceiling had been doubled from $1 billion to $2 billion total, which is $514.5 million more than the Task Force RIO and the Iraq Ministry of Oil had agreed was required in the Final Work Plan. The increase seemed excessive for repairs needed because of looting and sabotage since the plan was signed in late July, but it would keep bidders in the game. If the game went into overtime, it would benefit Halliburton KBR.

The Corps's amendment said "the subsequent contract"; the Army was still thinking in the singular. But what were bidders to do by November 5? They were to redo their responses to one of the two sample projects, the firefighting task, even though all nine well fires had been extinguished months ago. The original requirement had been to submit a detailed cost estimate and a narrative description of the bidder's plan for executing the work. "Even though the technical description of the sample project is abbreviated," the original RFP had said, "the cost proposal should represent the requirements for the full work effort..." It had said bidders should assume the work site was in Iraq or Kuwait and all types of facilities could be on fire — oil wells, gas/oil separation plants, pipelines, etc. The RFP had instructed bidders to describe and price "the capability to work one fire at a time and extinguish two fires in 10 days" and include "specifics of how they would provide firefighting water in the desert."

Although the amendment added some details, the technical description of the sample project was still extremely vague. The Corps wanted detailed plans and costs, but the bidders would interpret and respond to the vague requirements so differently that their bids could not be easily compared. Only one bidder would have the additional knowledge required to prepare a correct bid: Halliburton KBR. The amendment also split the sample project into two, one in the north and one in the south, with different parameters, so the bidders had twice the work to do in only seven days, compared to the 38 days they had originally had. And why even do it? What was the Corps's purpose is making the sample project a firefighting one? All of the well fires were out, and Halliburton KBR had already been paid millions of dollars to ready firefighting teams and equipment in Iraq in case saboteurs started more.

Back in early June, Dow Jones Newswire had quoted Gary Loew, Planning Director for Task Force RIO in Kuwait City, as saying there might not be enough time to award a second contract and meet the deadlines for completing the Corps's mission in Iraq.[12] If there was not enough time then, there was much less after the revised proposals were submitted November 5, unless the mission he had been referring to was not the same as the mission described to the bidders that hot Dallas day.

On December 2, the headlines in the Houston Chronicle read, "Iraq contract award delayed again, decisions probably slated for next year."[18] The Army Corps of Engineers in Fort Worth said it was postponing the award of the two new contracts to as late as January 17. The reason given? The holidays would give Corps personnel evaluating the bidders' proposals a few extra days — and Halliburton KBR more time to finish the work supposedly waiting for the bidders. By December 2, the value of the sole-source contract was up to $1.72 billion.

If the Army were to award two new contracts in late January 2004, it would have been a year since the Pentagon must have decided to award Halliburton KBR the sole-source contract. That contract was awarded March 8, 2003, and it would probably have taken six weeks to formulate and negotiate the contract terms and get the final contract approved and signed. The Army could have done it faster, but they were not in a hurry because it was done in secret – the contingency plan and the sole-source contract to implement the plan were both labeled "classified." If the Army were to award two new contracts in late January, it would have been ten months since the Corps of Engineers announced the sole-source contract on March 24, 2003, and promised to replace it by holding a "full and open competition." With that announcement, the media, the public watchdogs, and Representative Waxman intensified their charges of political cronyism and contractor corruption. The irony is that the more attention they focused on Cheney and Halliburton KBR during the rest of the year, the more the Army delayed the replacement contract awards, and the more money the company made.

Meanwhile, the furor continued over Halliburton's overcharging for fuel in Iraq. On December 10, The New York Times reported that KBR's price had increased to as much as $3.06 per gallon and the company would be paid an additional 5 percent if, in the judgment of the Corps of Engineers, it had "successfully administered the gas contract."[19] The next day, Dow Jones Newswires reported that, according to Pentagon auditors, Halliburton KBR had overcharged the Defense Department $61 million for delivering fuel in Iraq as part of its sole-source contract.[20] The report said Pentagon officials were concerned about "the problems with KBR's contracts, which were awarded without competitive bidding for up to $15.6 billion." The report quoted Dov Zakheim, Pentagon Comptroller, as saying, "Contractor improprieties and/or contract mischarging on departmental contracts will neither be condoned nor allowed to continue."

And what of government improprieties and contract mismanagement by the Corps? Back in October, a Corps spokesman, Scott Saunders, told the Washington Post that KBR was purchasing the fuel in Kuwait at the lowest price possible and that the three agencies monitoring the contract had found "no material errors."[21] Surely three government agencies hard at work protecting our tax dollars could not all be wrong.

On December 17, the Corps came to Halliburton KBR's defense, stating it had found nothing irregular in its own audit. Even Comptroller Zakheim claimed the company had not profited from the overcharge (but a Kuwaiti subcontractor certainly did and at U.S. taxpayers' expense). He said the overcharge may have stemmed from an antiquated accounting and cost-estimating system that "created problems for the company in reconciling some of the labor rates they have used as a basis for charges."[22] Now that is a very strange thing for the Pentagon's budget chief to say because U.S. Government contractors are only allowed to use systems certified by the Defense Contract Audit Agency (DCAA). Besides, until Zakheim got involved, the overcharge had been attributed to KBR's paying too high a price to buy the fuel in Kuwait, not to KBR's labor rates.

The next day, the finger pointing began. Halliburton claimed it had not overcharged the government but had actually saved the program $164 million by importing fuel from Turkey, as well as Kuwait. KBR chief executive Randy Harl said "the Army Corps of Engineers pushed the company to buy the fuel from Kuwait, where it had to pay a subcontractor $2.27 a gallon."[23]

In response, Corps spokesperson Bob Faletti admitted the Corps "might have" directed KBR to buy fuel in Kuwait initially because it was closer, and fuel was urgently needed to quash civil unrest. He still maintained that the Corps's own audits found no improprieties. [23]

The following day, Senator Joseph Lieberman, ranking member of the Senate Governmental Affairs Committee, jumped into the fray, alleging that Halliburton KBR's own internal auditors had warned management that they were overcharging for the fuel.[24] He said his charge was based on information from one of the DCAA auditors, who claimed to have seen the internal company report. In a letter to Defense Secretary Donald Rumsfeld, Senator Lieberman said the DCAA auditor discovered the report, took notes, and then confronted the company.

In response, a Halliburton spokesperson claimed that every page of the report was marked "proprietary and confidential for KBR internal use only," so the DCAA auditor violated the law; Halliburton refused to turn over the report.[24]

The Senator quoted DCAA Deputy Director Michael Thibault as saying Halliburton's own auditors had warned the company that it was charging excessive prices for the fuel.[25] He called for an investigation into Halliburton's "fitness to receive federal contracts" and possible suspension or debarment. According to the FAR, a contractor can be suspended for up to 18 months for a crime related to business integrity, failure to meet contract standards, or other serious misconduct. A contractor can be barred from receiving additional contracts for up to 3 years or more for a criminal indictment or conviction for fraud, embezzlement, or forgery. Either way, the contractor must show a pattern of conscious, consistent misconduct. Suspension is rare, and debarment is almost unheard of.[26] Fortunately, the FAR also provides for punishing misconduct by evaluating past performance when awarding new contracts. The Corps could have used Halliburton KBR's poor performance on its ongoing Iraq oil field contract, as well as its previous contract in the Balkans, to eliminate the company from receiving one of the two new contracts.

What of the U.S. Army Corps of Engineers' fitness to select and manage its contractors? The DCAA had identified the Iraq fuel overcharge in a routine contract audit, which ignited the uproar, but in the end, the DCAA will submit its final report to the Corps in Fort Worth, and it will be the Corps's responsibility to take whatever action it deems appropriate, if any. But to take action against Halliburton would have been courting disaster; more and more of the Corps's own misconduct might have been revealed.

And as long as the fuel overcharge continued to make the news, it was in the Corps's own best interests to delay awarding the new contracts. The Corps needed Halliburton KBR to be one of the two best qualified contractors in order to justify its original sole-source award. If it were not for the furor over the fuel, it would have been easy to award KBR yet another contract. According to Corps logic, the best qualified company is the company that did the preceding work. That was the Corps's explanation for awarding Halliburton KBR the $7 billion sole-source contract after it did the oil field contingency planning, as well as awarding the company that other follow-on, sole-source contract back in the Balkans after KBR lost the LOGCAP rebid to DynCorp. By the Corps's logic, no other contractor would have a chance to break the chain of awards to Halliburton KBR.

On New Year's Eve, the following headline appeared in the Washington Post: "Military Ends Halliburton Deal to Supply Gasoline to Iraq." The Pentagon had announced that it was relieving Halliburton of its responsibility for importing and delivering fuel for domestic consumption in Iraq and turning the task over to the Defense Energy Support Center, which had been doing the same thing for the armed forces for only $1.19 per gallon. However, the headline was a bit premature. According to Pentagon officials, the Center would not actually take over the task from Halliburton for several months because it needed 60 to 90 days to solicit and award new fuel subcontracts.[27]

With the new year came new revelations. It was discovered that, back on December 19, the Army had waived the requirement for Halliburton to provide cost data to justify the high prices it charged to deliver fuel in Iraq.[28] The FAR allows a cost justification waiver under exceptional circumstances when there is evidence, other than cost data, to demonstrate that the prices charged are fair and reasonable. The waiver the Corps provided Halliburton stated that the company "has been unable to secure any real competition for these services." The waiver contradicted itself by also stating that "other trucking and oil distributors have made proposals to [Halliburton] on more favorable terms" than the Kuwaiti subcontractor selected.[29]

Halliburton was using the mysterious Altanmia Commercial Marketing Company, a Kuwaiti entity with no fuel transport experience. Petroleum industry insiders in the Middle East had never heard of Altanmia. Although other Kuwaiti companies offered to deliver the fuel at lower costs, Altanmia was the only company the Kuwaiti government would license to perform the work.[29] Owned by a prominent Kuwaiti family with possible ties to high-ranking Kuwaiti government officials, records indicate that Altanmia is involved in unspecified work for the Kuwaiti military, as well as various real estate ventures.

The waiver, in effect, exonerated Halliburton KBR of any wrongdoing in the fuel overcharge matter, overcharges that totaled $61 million at the end of September and were still accumulating at the rate of about $20 million a month. According to Representative Waxman, the Commander of the Army Corps of Engineers, Lieutenant General Robert Flowers, signed the waiver at Halliburton's request just nine days after the DCAA accused the company of refusing to turn over the internal audit report seen by the DCAA auditor — and just four days after The Wall Street Journal reported that the Corps contracting officer responsible for overseeing Halliburton's contract had complained of "political pressure" she had received from both the Bush administration, via U.S. Embassy officials in Kuwait City, and the government of Kuwait, via the state-owned Kuwait Petroleum Company, to deal only with Altanmia.

Representative Waxman described General Flowers' action as "deliberately sabotaging the government's ability to audit Halliburton."[29] The waiver was retroactive and applied to other Halliburton subcontractors' costs, not just Altanmia's. In effect, the waiver eliminated the chances of any meaningful audit in the matter. In the words of Representative Waxman, the waiver was a whitewash.[30]

Back on December 2, the Corps had promised to award the two new Iraq oil field contracts no later than January 17. But on January 13, General Flowers was quoted in the Washington Post as saying the awards had been postponed yet again: "When you have allegations, you have to see to things before you proceed."[31]

"We can't bar KBR," he said. "We took pains so there would be a level playing field, so there wouldn't be a marked advantage. We can't deny someone who is by law free to compete."

So when the playing field needed leveling to benefit Halliburton KBR, the Corps was apparently ready, willing, and able to do that. It had not been willing to open the playing field for the oil field contingency planning when the advantage might have been Bechtel's. It had not been willing to open the playing field for the $7 billion Iraq oil field contracted awarded secretly to KBR. When the playing field was finally opened for the two new contracts, it turned out to be a field of dreams.

No sooner than the DCAA asked the Pentagon's Inspector General to conduct a formal investigation of irregularities in Halliburton's books and "evidence of potential unlawful activity," General Flowers announced that the new contract awards would be delayed until March.[32] As he said before, "When you have allegations, you have to see to things before you proceed."

The ink was not yet dry on that announcement when the Corps reversed itself. On January 16, they awarded Halliburton KBR the larger of the two new contracts, valued at $1.2 billion, for work in the southern oil fields in Iraq. The other new contract, valued at $800 million for work in the northern oil fields, was awarded to a joint venture between Parsons and the Worley Group of Australia. Halliburton's award brought the total value of its Iraq contracts to over $10 billion. Halliburton's CEO called the award "validation," while the Waxman word for it was "mind-boggling."

"It's special treatment to reward the company with yet another contract in the face of all these unresolved questions," Representative Waxman said.[33]

And it's special treatment to reward the Army Corps of Engineers for its misdeeds by laying all the blame for the fuel overcharges on Halliburton KBR. The Corps's contracting officer and staff was responsible for overseeing KBR, ensuring that the company complied with the law, and making certain that taxpayers' dollars were well spent. When a minor commits a crime or other heinous act, society does not relieve the parents or guardians of all responsibility for their charge's egregious behavior. When a child uses a careless parent's gun to commit murder, society prosecutes the parent, too.

On January 23, the news broke that at least one Halliburton employee had accepted a $6.3 million kickback from a Kuwaiti company in return for a subcontract to provide logistics support for U.S. military troops in Iraq, part of Halliburton KBR's LOGCAP contract with the Army. [34] Apparently, Halliburton had informed the Defense Department of the kickoffs in a letter to the Inspector General dated January 15, the day before the new Iraq contract award![35]

"It is incomprehensible that the administration could give Halliburton another billion-dollar contract without fully investigating such serious criminal wrongdoing," Representative Waxman wrote to the chairman of the Committee on Government Reform, Representative Tom Davis (R-Va.), in calling for a series of hearings on waste, fraud, and abuse in the Iraq reconstruction contracts.[35]

And it's incomprehensible that the Army Corps of Engineers awarded the new contracts when it did, for any reason other than to benefit Halliburton KBR. The delay until March would have allowed time for the fuel overcharge matter to be resolved, or the focus on it fade. The delay until March would have enabled KBR to complete even more work under the $7 billion sole-source contract before there could be any question of transitioning the work in Iraq's northern oil fields to the Parsons team. But Halliburton's letter to the Inspector General triggered damage control. The new contracts had to be awarded before the news of the kickbacks leaked out, before any hearings convened, before the draft DCAA report on the fuel overcharges became final, and before any more improprieties were revealed.

As January turned into February, new overcharges were found by DCAA auditors examining Halliburton KBR's LOGCAP books, not by the Army staff responsible for overseeing the contract. KBR had charged the government $27.4 million for meals never served to troops in Kuwait and Iraq.[36]

8

How to Start a Revolution: Eliminate the Smoke and Mirrors

THE ENDLESS DELAYS in the award of the two new Iraq oil field contracts made them obsolete, or should have. According to the *Restoration of Iraqi Oil Infrastructure Final Work Plan*, the remaining work needed to return Iraq's oil fields to pre-war production capacity had been agreed upon in July 2003 by the U.S. Army Corps of Engineers, the Coalition Provisional Authority, and the Iraqi Ministry of Oil (MOO). According to the Final Work Plan, all of the necessary work, and the funding for it, had been committed to either Halliburton KBR or directly to the MOO, and all of the work was scheduled to be completed by the end of March 2004. Whether Parsons will get any work out of its new contract, awarded in January 2004, remains to be seen. According to the Corps, it would probably take until March for Parsons' oil field team to mobilize and become operational in Iraq.[37]

In late January, the MOO announced its success increasing production of crude to more than 2.3 million barrels a day from almost nothing when the war ended. The southern oil fields were producing at near capacity, and crude from the northern fields was expected to start flowing to overseas markets within days. Security guards had been trained and stationed along the pipeline from Kirkuk to the Turkish port of Ceyhan on the Mediterranean, in order to protect it from saboteurs, like those who bombed it repeatedly, closing it completely the previous year. The MOO announced that it expected to surpass the pre-war production capacity of 2.5 million barrels a day by the end of March.[38] If the U.S. Government's commitment to Iraq would in all likelihood be met by the end of March or shortly thereafter, why were we committing $2 billion for extra work by either Halliburton or the Parsons team? If the House Committee on Government Reform holds the hearings requested by Representative Waxman on waste, fraud, and abuse in the Iraq reconstruction contracts, beginning with Halliburton, and if the Corps personnel guilty of deception in Dallas and other deceits are called to testify, and if they testify truthfully, perhaps they will tell us.

But more likely they will not. It is more probable that these Corps personnel and their accomplices will remain uncensored, because big business will continue to be demonized instead. The critics of Halliburton and Bechtel apparently do not understand how U.S. Government contracting works, so they simply blame the 'evil' corporations if anything goes awry. The laws and regulations that govern the actions of federal contractors and their employees — including laws against fraud and abuse — apply to their government "handlers" as well. Federal contractors do not operate in a void; they are subject to extensive oversight. Every government agency has an entire bureaucracy devoted to administering contracts for goods and services. Every cost-reimbursable, ID/IQ contract requires extensive agency review/approval cycles for all of the contractor's proposed project plans and task order cost estimates before work begins, as well as a steady stream of cost/schedule status reports once it does. If a contractor is dramatically overcharging the government, the contracting officer should know about it long before it is uncovered by a DCAA auditor.

Despite the mounting evidence of procurement fraud, political favoritism, and contracting abuse by the U.S. Army Corps of Engineers, going back years (remember the Balkans?), Iraq contracting critics both in and outside government remain focused on Halliburton KBR. The federal government appears incapable of any real self-examination despite — or perhaps because of — the mounting evidence that fraud and abuse were committed within their own ranks. There is a growing body of evidence that the Army has grossly mismanaged a number of cost-reimbursable contracts for logistics support abroad, including but not limited to, putting its stamp of approval on outrageous contractor prices for supplying plywood in the Balkans and fuel in Iraq. Nevertheless, the critics continue to focus on the contractor, while the Army's contract management personnel — the individuals with front-line responsibility for spending taxpayers' dollars wisely by monitoring contractor performance — escape with no demerits. We trust mutual fund managers with our retirement dollars, but when they show a pattern over time of failing to monitor the performance of the companies in which they invest our money, and we lose it, they do not escape unscathed. Nor do inside traders, if caught. Why should federal agencies and employees be protected from both the law and market forces when they fail in their fiduciary duties to safeguard our investments, be they in America or Iraq?

Contractors' fraudulently deceiving the government is considered a crime. Apparently, the government's fraudulently deceiving contractors is not. What action, if any, will be taken against the Corps personnel who intentionally misled Halliburton's competitors at the bidders meeting in Dallas on July 14, 2003? What action will be taken against the Corps personnel who withheld the *Restoration of Iraqi Oil Infrastructure Final Work Plan* until shortly before the bidders' proposals were due? Most of the bidders would have dropped out after the meeting and saved hundreds of thousands of dollars in proposal preparation costs if the Corps had been honest, told them at the meeting about the workshop held in Baghdad the previous week, and revealed the contents of the Final Work Plan developed there. When it was finally released and bidders realized the work had already been committed to Halliburton or the MOO, why did they not complain? Contractors will not speak negatively (publicly) of their customers for fear of reprisal; they will not jeopardize their chances for future U.S. Government work. They are, in effect, coerced into abiding by a code of silence; according to the rules of what amounts to a secret government/industry fraternity, the successful federal contractor does not squeal. Why didn't the rest of the bidders drop out and cut their losses, as Bechtel had done? For much the same reason; the Corps needed a "competition" to take place.

The critics and the media had a field day accusing the Bush administration of political cronyism in the award of Iraq reconstruction contracts to both Halliburton and Bechtel. It started in March 2003, when Halliburton KBR's sole-source contract was first revealed, and continued for much of the year until the spotlight turned to other events, including the capture of both Saddam Hussein and the Democratic presidential nomination. The charges of political cronyism resurfaced occasionally when new revelations appeared, such as the $6.3 million kickbacks accepted by Halliburton KBR personnel. But the firestorm was waning, and the whole Iraq contracting debacle was fast becoming old news. That was unfortunate indeed because it could have triggered major introspection by government agencies and much-needed federal contracting reforms.

"Turf Wars: Real combat, plastic men, and they're fighting in your backyard... for 1 to 4 players" — video game ad

Even before contractors began vying for the opportunities that came with the reconstruction of Iraq, a turf war was waged within the U.S. Government. From the start, the Defense Department and USAID had lobbied for control of the rebuilding funds, and the administration's solution was to divide the spoils, with the Iraq oil field work going to the Army Corps of Engineers and the country's civil infrastructure reconstruction given to USAID. Other agencies fought over lesser territory than the capital reconstruction program. In anticipation of the second round of funding, the turf wars resumed, along with the accusations in the press against both the Pentagon and USAID for lack of planning, political cronyism, and slow progress on the ground. Construction had always been the Corps's mission, but USAID had shifted its focus to humanitarian efforts around the world and had not been involved in major capital projects in many years, certainly none the size of those needed in Iraq. In late May 2003, not long after Bechtel and other USAID contractors began mobilizing to

Iraq, the Agency "hired" the Corps of Engineers to provide technical advice and ensure the contractors' work was completed in accordance with USAID's job orders and U.S. Government policies. The cost of the Corps's services was set at $3 million a year.

This interagency relationship should have facilitated the reconstruction in Iraq, but it did not. For example, Bechtel was heavily criticized in the press for less than rapid progress restoring the electricity in Baghdad and other cities. Part of the problem was looting and sabotage. Hundreds of transmission towers were blown up by saboteurs, and the transmission lines disappeared into smelters almost as fast as they could be replaced, due to lack of security. The other part of the problem, less publicized, was the lack of natural gas and fuel oil to run the power plants that Bechtel's team had repaired because of problems restoring Iraq's oil fields. Halliburton was criticized for slow progress there, but the oil wells and refineries had input requirements beyond their control as well — they needed electricity to operate. It was the proverbial chicken-or-an-egg dilemma. With USAID and the Corps working together, one would have thought the two programs would have been coordinated in order to optimize the use of critical resources for the greatest good.

Rather than require the Army Corps of Engineers, USAID, and their contractors to coordinate their agendas, interagency integration was assigned to the newly created Coalition Provisional Authority (CPA), which replaced the short-lived Office of Reconstruction and Humanitarian Assistance. Was the CPA effective? Newsweek wanted to know, for example, why the Daura power plant, the largest in Baghdad, was still off line in October 2003, more than five months after President Bush declared an end to the war and nearly six months after USAID awarded the civil infrastructure reconstruction contract to Bechtel.

When interviewed in November, USAID Administrator Andrew Natsios "referred to a priority list drawn up by a coordinating committee under the Coalition Provisional Authority — the chief occupying power — and said he didn't know where Daura was on it."

When an aide interjected that the CPA would know, Natsios said, "No, Bechtel would know."

But Bechtel's spokesman told Newsweek, "We perform the work tasked to us by USAID. We don't make decisions on priorities. USAID and CPA make those decisions."[39]

Newsweek found that, once reconstruction decisions were finally made, most U.S. officials failed to follow up with adequate oversight of the contractors responsible for the work. A USAID staffer went as far as to compare the old (Hussein) and new (CPA) regimes. "Saddam had better accountability," the staffer said. Many of the contractors agreed, complaining that the CPA employees overseeing the reconstruction work were mostly political appointees who did not understand the process. "Every time we turn around there's a new order coming from CPA, 'Do it this way; no, do it that way instead.' It's just unbelievable."[39]

The OW! Factor

How had USAID and the CPA attempted to remedy the oversight problems? Back on May 25, 2003, USAID had contracted for the Corps's services to assist its 14 employees in Iraq in overseeing more than $1.5 billion worth of work under Bechtel's contract and nine others. Exactly one month later, USAID awarded yet another contract, this one to Management Services International, for assistance in monitoring the ten reconstruction contracts. The additional cost to taxpayers was $15 million. So USAID's 14-person staff had yet another contractor to manage, while moving a step farther away organizationally from direct oversight and accountability for the reconstruction work. USAID's OW! factor was 5; it needed one management contractor to provide oversight (O) for every five prime contractors performing reconstruction work (W), i.e., work yielding tangible progress toward completing the Agency's mission in Iraq.

Meanwhile, the Pentagon was taking a similar approach in its attempt to improve oversight and coordination. In order to answer the critics calling for greater transparency and accountability for how taxpayers' dollars were being spent. The Defense Department did what the government always does. It created yet another layer of bureaucracy by setting up a completely new program management organization and interjecting it into the hierarchy between Ambassador Bremer's CPA staff and the agencies closer to both the oil field and civil infrastructure work, including USAID and the Army Corps of Engineers. The Pentagon announced the formation of the CPA's new Program Management Office (PMO) on November 19, 2003.[40] It would be located in Baghdad with a staff of 130 from various federal agencies, and it would be headed by Rear Admiral David Nash, U.S. Navy, Retired. The PMO would award 24 new contracts — 13 of them worth $1 billion each.

Seven contracts would be for program management and oversight office services. One of the seven contractors would provide overarching PMO support by coordinating and integrating the reconstruction work in all six sectors: (1) Transportation and Communications; (2) Electricity; (3) Buildings and Health; (4) Security and Justice; (5) Public Works and Water Resources; and (6) Oil. This top management contractor would assist the PMO in overseeing the six other management contractors, one for each new Sector Project Management Office (SPMO). These six mid-level management contractors would in turn oversee 17 other prime contracts awarded by the PMO for sector- or sub-sector-specific construction. These 17 contractors would more than likely parcel out the actual work to subcontractors, including Iraqi firms. Together, there would be about 2,000 projects to complete.

So the new PMO would have two OW! factors, 6 (i.e., 6/1) and 3 (i.e., 17/6 = 2.833, rounded up to 3). In mathematics, the factorial of a number n — written $n!$ — is the number of ways in which n objects or entities can be permutated, or rearranged. For example, USAID's 5! = 5 x 4 x 3 x 2 x 1 = 120. As n increases slowly, $n!$ increases dramatically: the PMO's top-level 6! = 720 different permutations! So, the OW! factor is one way of representing the complexity of any given program management organization (aka number of potential headaches for the agency at the top) and predicting what will happen if the number of management contractors increases. The PMO's lower-level 3! = 3 x 2 x 1 = 6. With two levels of management contractors interacting, the permutations must be multiplied: 720 x 6 = 4,320 potential problems or conflicts among the primes. (The prime contractors' potential problems with their subcontractors can be similarly predicted.)

With the PMO, the government had created a brand new bureaucracy in which contractors would oversee other contractors overseeing still other contractors, all working under cost-reimbursable terms — a pyramid scheme if there ever was one, and like all pyramid schemes, bound to collapse eventually. The terms included an automatic fee equal to 3% of all costs billed to the government and another percentage fee for exceptional performance, which would presumably be assessed by the contractor at the next level up, not the government, until one reached the very top.

Even with a staff of 130 government employees and seven management support contractors, Admiral Nash's job would be far from simple, for all of the new contracts would be cost-reimbursable, like the Halliburton and Bechtel contracts that had caused so much stir. Each contractor, including the seven overseers, would receive an award fee equal to 3 percent of the total contract cost; good performance would increase the total fee by additional percentage points. The contracts would be awarded by January 31, 2004, using an accelerated process for "full and open competition."[40] Such competition usually takes most well-established federal agencies at least six months, although USAID did it in half that time twice, when the Agency selected and awarded its two largest Iraq civil infrastructure contracts to Bechtel.

"It's very fast for our system," said Admiral Nash, "but we're going to do it all appropriately and make sure we comply with all the regulations and law. To me, it's not impossible. It does make people wonder. But I think we're okay."[40]

In less than a month, the 24 contracts had grown by two, and the Army Corps of Engineers had taken over more than half of them from Admiral Nash's fledgling organization.[41] On December 10, all 26 procurements were put on hold, while the infighting within the Bush administration continued. Admiral Nash resigned and left Baghdad with his senior staff to protest the decentralized oversight and accountability that would result from the Corps's destruction of the unified PMO plan.

"I personally feel that we are setting ourselves up for failure," he told Ambassador Bremer.[42]

In early January 2004, the administration announced that contracts for much of the reconstruction to be funded by the additional $18.6 billion Congress approved the previous fall would not be awarded until after the planned interim Iraqi government takeover on June 30.[42] Admiral Nash and his staff agreed to return to Iraq, but their responsibility was reduced to awarding and overseeing only $5 billion worth of work, while $2 billion went to USAID. The fractious Army Corps of Engineers ended up having to share the remaining $1 billion in near-term funds with the Air Force, so perhaps there was some justice after all. Perhaps someone, somewhere in all those layers of management bureaucracy, was beginning to wake up to the problems at the Corps. On the other hand, the much needed centralized coordination and accountability were gone.

"The significant problems of mankind will not be solved until we advance beyond the level of thinking we had when we created them." — Albert Einstein

The federal government's response to calls for better contract oversight is ironic. Government officials have responded to the demands for greater transparency and accountability for how taxpayers' dollars are spent, not by rolling up their sleeves and diving into the day-to-day details in order to route out the problems, waste, and abuse, but by removing themselves even further. The CPA's new PMO organization would add layer upon layer of management structure. Like water, the deeper the bureaucracy, the less transparency. It's harder to see what's happening at the bottom, because it's farther away and much of the light that does reflect up is absorbed, deflected, or distorted by intervening matter, in this case bureaucrats. The deeper the organization, the higher the management costs, so there is a smaller percentage of the total program funding available to accomplish the work that is the program's purpose, in this case, the reconstruction of Iraq. We criticize charities and other nonprofit organizations when salaries and other administrative costs take too large a bite out of our donations and too little goes to the needy or the good work for which our donations were intended. Why shouldn't the government be called to task for the same lack of financial stewardship of our tax dollars?

Obviously, bureaucracies are unavoidable, but government needs to learn a lesson from the private sector, where extremely lean organizations have become the norm. Despite significant downsizing, government agencies have remained rigidly stove-piped, while contractors have adapted to lower profit margins by dual-hatting their personnel to field the smallest program management teams possible without jeopardizing effective oversight of the contract work. Cost-conscious private industry drives decision-making and accountability down to the lowest possible level, but in government, it is reversed. Low-level civil servants tend to only analyze information and make recommendations, driving decision-making and accountability upward. As a result, government cannot act or respond to changing requirements or conditions as fast as the private sector, where the decision-makers are more hands on.

Government agencies are more likely to get stuck in mental ruts, unable to recognize either scattered clues or long-term trends that suggest something is going wrong, that their traditional approaches are not working. Private companies cannot afford to have their heads in the sand; their very existence depends on constant vigilance and self-examination, continual readjustment, and occasionally massive retooling or reorganization to stay competitive, efficient, and profitable. Because most companies are open to new ways of doing business, technologically or otherwise, and their organizations are running leaner, their employees no longer have the job security they once took for granted. Not so in government. Military or civil servants who oversee federal contractors are less concerned with identifying waste and abuse because their jobs do not depend on it.

For the taxpayer, asking the government to improve accountability is a double-edged sword, because the only approach federal bureaucrats can apparently think of — expanding the bureaucracy — increases program costs with no guarantee of results. It is simply smoke and mirrors. The approach is counter-productive because it does not bring the positions of ultimate accountability any closer to the execution of the work. Instead, adding layers of management, as in the complex multi-tiered PMO, lengthens the chain of command and associated communications channels, making fact-finding, decision-making, feedback, and work progress slower, more difficult, and more costly. When are federal agencies going to learn to work smarter with what they've got?

To ensure that taxpayers' dollars are well spent, federal agencies rely on extensive cost and progress reporting by contractors and occasional inspections and audits by government staff. Contract oversight should consist of more than identifying cost overruns after the fact. It should be proactive instead of reactive, more real-time instead of post time. In a word, it should be "interactive," but the government's approach will not change until its basic mentality changes. The government assumes contractors are inherently greedy and amoral and have no pride of workmanship; they will lie and cheat as much as they can get away with, unless government auditors and inspectors stop them. Federal agencies are the Forces of Light, and contractors are the Forces of Darkness. In response, contractors naturally become more secretive, even if they have nothing to hide, and scorn their overseers.

Without mutual respect and open communication between government and industry, valuable information sharing and feedback are lost in both directions. The loss is especially great on large, complex projects and cost-reimbursable contracts with many unknowns, like those in Iraq. How can the government's contracting officers know for sure whether Halliburton and Bechtel performed their assigned tasks as well as humanly possible under the particular circumstances of time and place? They can't — unless they are embedded with the contractors' project teams. "Interactive oversight" could be provided by embedded contracting officers and staff with the contractor's project personnel in the contractor's project office on a continuing basis. This may be a revolutionary idea to federal agencies, but contractors have been doing it themselves for years. In Russia, for example, U.S. contractors use Russian subcontractors to perform Cooperative Threat Reduction work under contracts with the Defense Threat Reduction Agency, part of the U.S. Department of Defense. Best commercial practices, taken for granted in the United States and Western Europe, are not yet the norm in Russia, where most management and accounting procedures reflect their Soviet-era origins. For example, Russian cost estimates are more often prepared top down (what a subcontractor thinks the prime contractor will pay), rather than bottom up (what the work actually costs plus a fair profit). As a result, U.S. contractors have embedded subcontract management and technical personnel in Russian subcontractors' organizations, with the number of oversight personnel depending on the amount and importance of the subcontracted work, as well as the skills of the subcontractor's staff. In addition to giving the prime contractor real-time input to and feedback on the subcontractor's performance, a more skilled and efficient workforce is left behind when the U.S. contractor's work in Russia is done. It's a win-win situation for all concerned.

With fewer and fewer contract management personnel in the federal government, as described in the Preface, and more and more work being contracted out to private industry, it's time for the U.S. Government to consider new approaches to contractor oversight. Interactive oversight by embedding federal contracting officers and staff in contractors' organizations should be considered, not for every contract, but for the large, complex, cost-reimbursable ones, especially considering the ongoing trend in which many once-separate small contracts are bundled into fewer larger ones. Agencies can pay their contract oversight staff to report to the contractors' project offices as easily as they can pay them to work in federal offices in Washington, D.C., Alexandria, Virginia, or elsewhere.

On U.S. Government projects abroad, contractors are often required to provide office space, equipment, and support services for the contracting officers and staff when they are in country, such as for inspections or quarterly review meetings. Even in Iraq, where the Army Corps of Engineers and USAID have contracting personnel on full-time assignment, they are not functioning as interactive oversight teams working side by side with contractor personnel on a daily basis. Even when contractors are required to provide government staff with real-time access to the cost/schedule control and accounting systems they are using on federal projects, they can limit that access to the data they wish to reveal by password-protecting the rest.

Contractors will undoubtedly resist interactive oversight if it means opening, not just their books, but every closed-door discussion that pertains to the publicly funded program. However, their right to privacy should end when taxpayer dollars are paying their employees to attend. Although the very large federal contractors already provide permanent offices for DCAA auditors in their corporate headquarters, the auditors are still auditors after the fact. They examine financial records; they are not embedded in the contractors' project organizations, they do not participate in managing the work from day to day, and they do not experience first-hand the same pressures or extenuating circumstances that affect contractors' management decisions.

Auditors file reports presenting their findings; the responsibility for taking action belongs to someone else. When the DCAA auditors submit their final report on the Halliburton fuel overcharges in Iraq to the Corps of Engineers, the auditors are done. The Corps has the power to file it away, justifying the $61 million with circumstances after the war, and if other irregularities had not come to light, the Corps may have been right. In theory, giving contracting officers enforcement power should work, and in practice, if they are doing their jobs, it does. The actions and statements of Corps personnel, presented in Chapter 7, suggest that, instead of doing their jobs, they were either in bed with Halliburton or asleep at the wheel.

Government agencies and contractors often launch new projects with "partnering" sessions, but more often than not, the cooperative spirit soon wanes, as the two sides assume their adversarial positions and differ over change orders and other contractual issues. With the proper paradigm and guidelines for real-time, in-house interactive oversight, in which the government's contracting officers and staff work side by side with the contractor's project management personnel, perhaps the finger-pointing and scapegoating could be avoided, or at least dramatically reduced, and the economic risks for taxpayers would be minimized. The federal contract overseers would not be caught unaware, or at least they could not use that excuse. They would know about a problem and be able to take action immediately, not long after the fact. If the military or civil servants filling the contract oversight positions were held as accountable for cost and performance as are contractor personnel, the two forces would have common goals, and a win-win situation would be created.

What If?

With the intermingling of government and contractor personnel, the way in which work is planned and executed would trend away from the government's traditionally linear mode of thinking and organizing itself to industry's more decentralized approach. The government typically decides on a desired outcome and plots the steps required to realize that end. The resulting strategic plan would only work if the government's intelligence and projections were accurate and there were no significant surprises. Strategic planning is not very effective in an unpredictable world like ours post 9/11.

As Philip Bobbitt, University of Texas law professor and senior director at the National Security Council during the Clinton administration, put it, "Governments sought the likeliest linear future and planned accordingly. Unfortunately, in an increasingly decentralized world, in which previously insignificant actors and factors can play a decisive role, strategic planning can leave decision-makers flat-footed. In its uni-dimensional reliance on a single future, strategic planning hardens the 'official future' agencies internalize, and thus prepares them poorly for appreciating rapid changes in circumstance and for making agile adaptations."[43]

Strategic planning has been abandoned by much of private industry — and was abandoned by most of the Clinton administration — in favor of scenario planning (although companies may still refer to what they do as "strategic planning"). Companies analyze numerous what-if scenarios for potential impacts on a proposed program and then evaluate a wide variety of alternative responses. They often use computer modeling to help develop and assess alternatives based on the many known variables in a complex situation and the full spectrum of decisions that could be made by any of the players in the game, not just the desired decisions and one's own.

According to Professor Bobbitt, "Getting the government to emphasize scenario planning will not be easy. To be successful, the approach depends on well-organized dialogue between decision makers at many levels, which would be culture shock for the rigidly hierarchical executive branch... Our various national security agencies may not be competitive businesses, but they often behave toward each other as if they were. Intelligence is often 'stove-piped' — when analysts refuse to share information and sources with interagency rivals working on the same problems — and mutually distrustful cultures abound."[43]

Professor Bobbitt was writing about the National Security Council being caught by surprise by a number of events predating September 11 and since. His assessment applies equally well to the turf wars fought by federal agencies for control of the Iraq reconstruction program. More broadly, it applies to virtually every relationship between a government agency and a contractor because of the inherent mutual distrust. "Scenario planning requires a political culture that is tolerant of uncertainty... competing scenarios are anathema to bureaucrats whose careers are threatened by answering questions like, 'What would it take for this estimate to be dramatically wrong?' — which translates to, 'What arguments can you give me that undermine your own recommendations?'"[43]

"Traditional governments are so preoccupied with rowing that they forget to steer." — David Osborne and Ted Gaebler

Given our uncertain world, cost-reimbursable, ID/IQ contracts are here to stay. They are awarded when there is uncertainty as to the work that will be required — where, what, for how long, and how much it will cost. Such contracts are inherently difficult for both the contractor and the government, but more so for the government — and the taxpayers footing the bills — if its contract oversight staff continue to sit at their desks, relying on an orderly stream of cost and progress reports to keep them informed. If the contract is also performance-based, the contractor is paid extra for exceptional performance, with the additional fee calculated as a percentage of the contract cost (as a result, a contractor is doubly rewarded for undetected overcharging).

Measuring contractor performance is as much a problem for the government as it has been for private sector managers assessing the performance of their employees. Managers at the Defense Department and Homeland Security Department have begun grappling with the problem as well, because the Bush administration and Congress are requiring them to implement the pay-for-performance approach to salary increases.[44] Despite some complex metrics that have been tried, the science of performance measurement is in its infancy. On large, performance-based, cost-reimbursable contracts like Halliburton's and Bechtel's in Iraq, there is no substitute for being there. The contracting officers must be more directly involved, day in and day, with the contractors they are overseeing, and must be held accountable for the contractors' performance, at the cost of the contracting officers' jobs if necessary. If a contracting officer's performance were directly linked to the contractor's performance, the attitude of "we're in this together" would soon evolve, and the resulting synergies would create a win-win situation for all concerned, including the American taxpayer.

Federal procurement reforms in the 1990s started a movement toward treating contractors as partners, not opponents to be held at arm's length, but it fizzled because of lack of structure and because contracting officers were not held accountable for its failure. Some would say the partnering movement not only failed but backfired. "The environment became one that encourages and rewards carelessness," according to Angela Styles, until recently the top administrator at the Office of Federal Procurement Policy. "The lines became too easy to cross, and no one was paying attention. I don't even think most people know where the lines are anymore."[45]

Federal procurement (selecting contractors) has evolved further than contract management (overseeing them) in realizing that, in contracting for services, the government is buying a project team — people, not widgets, who function well together. After bidders' proposals are evaluated and the list of competitors is reduced to a final few, most federal agencies interview the short-listed contractors' proposed project managers and other key personnel. Of course, the interviewees are on their best behavior and have been extensively prepped and rehearsed. Perhaps a better way for the selection board to determine how the personnel would actually function under real-world pressures and conditions would be through contracting officers who had been embedded with them. There are thousands of pages in the FAR, the Defense Department's supplement to the FAR, and the Army's supplement to that. No matter how many more laws and regulations are created in attempts to make federal contracting fair and honest, ensuring fairness in the selection of contractors, as well as honesty in their delivery of goods, services, and invoices for same, will always involve human judgment to some degree. Good judgment, especially under extraordinary circumstances, cannot be promulgated like regulations and procedures.

If the goal is to eliminate overcharging and poor performance by contractors, the government must make contracting officers fully accountable for same and must maximize their ability to be successful by embedding them with the contractors they oversee. Unfortunately, the solution to every problem has within it the seeds of more problems. The potential for abuse in interactive oversight is obvious, so a system of checks and balances would be required. Highly skilled contracting officers from various federal agencies could be consolidated into an elite interagency pool and deployed by an existing agency, such as the GAO or Office of Procurement Policy, not a new one. Individual officers could be assigned to, and rotated periodically among, the government's largest cost-reimbursable contracts. Best commercial practices could be used to attract and retain top-notch contracting professionals, including remuneration commensurate with performance, rather than tenure, and bonuses for completing projects under budget or ahead of schedule.

This interagency approach would facilitate the sharing of lessons learned between federal agencies, including more meaningful past performance information on contractors than is contained in existing databases. An interagency pool of talented contracting officers would help break down departmental stove-piping and end the turf wars within the government. It would foster integrated scenario planning, rather than unilateral strategic planning. Such a pool would have been the perfect place to have gone for assistance in planning and executing the Iraq reconstruction program. A pool of seasoned contracting officers, who had already worked together under fire on other high-visibility programs, would not have stumbled so badly in either awarding or overseeing the contracts.

The 1991 Defense Authorization Act mandated the creation of a high-level panel to review the FAR for applicability to Defense Department acquisitions. Known as the Section 800 Panel, its members came from both government and private industry. The panel issued its report in January 1993, recommending changing or eliminating nearly 300 regulations in order to streamline procurement processes and costs. The panel report, combined with Vice President Gore's National Performance Review later that year, led to the 1994 Federal Acquisition Streamlining Act.

In 1999, another high-level panel was created to review and improve the FAR and agency-specific acquisition regulations that implement or supplement the FAR. Its priorities included "creating an acquisition workforce of mission-focused business leaders" and "transforming the Federal Acquisition System for better business results."[46] Known as the Procurement Executives Council, its members came from various federal agencies; business leaders from private industry were excluded. In 2003, the Procurement Executives Council changed its name to the Federal Acquisition Council in order "to provide greater flexibility and a more inclusive reach beyond procurement." If the Council has produced any "mission-focused business leaders," they were not at the bidders meeting in Dallas, and they were not managing Halliburton contracts.

Just as the Defense Department created yet another bureaucracy to take over the contracting in Iraq, the government created yet another panel to take on the megalith that federal acquisitions improvement has become. In February 2004, Representative Tom Davis (R-Va.) created the new panel by adding language to the Fiscal Year 2004 Defense Authorization Act. The panel will have one year to study the problems, especially performance-based contracting, and to evaluate best commercial practices that could be adopted by the government, such as share-in-savings contracts. Then the panel will report to Congress. Hopefully, it will include business leaders from private industry, as well as government. At press time, the members of the panel were yet to be announced; they must be appointed by the head of the Office of Federal Procurement Policy, a post currently vacant. According to Angela Styles, last to fill the post, "Tom Davis is choosing the members, so he will likely dictate the results."[47] Other critics of the powerful Republican Congressman fear his panel's agenda will coincide with that of powerful federal contractors, like Halliburton and Bechtel.

After the new panel reports to Congress, it will be up to the executive branch to act on the findings, and the Bush administration can do one of two things. It can discard the findings and allow the government's own demons to go unchecked, as the Army Corps of Engineers did with the DCAA auditors' report on the Halliburton fuel overcharges, or it can embrace the findings, acknowledge what transpired in the oil field contracting for Iraq, and advance beyond that level of thinking. In either case, General Dwight D. Eisenhower's final speech as President of the United States in 1961 contains good advice: "Only an alert and knowledgeable citizenry can compel the proper meshing of the huge industrial and military machinery of defense with our peaceful methods and goals, so that security and liberty may prosper together."[48]

AFTERWORD

Perhaps there was never any political cronyism or other procurement fraud in the Iraq oil field contracting activities conducted by the U.S. Army Corps of Engineers' Fort Worth District — no intent to delay the "free and open competition" to ensure billions of dollars' worth of work would already be committed to Halliburton KBR, no intent to lie to the bidders in Dallas about the opportunities left for them, and no intent to withhold time-critical information that KBR had and they were entitled to as well. Perhaps there was never any bureaucratic chain of favoritism linking Vice President Cheney to the billions of dollars' worth of work secretly awarded to Halliburton KBR. Perhaps the only chain leading to those awards was one in which the links were simply act, after act, after act, of mismanagement and failure to comply with the law. I hope not. I'd rather believe it was political cronyism than believe such incompetence and poor judgment are so pervasive in our government that they are no longer recognizable as such.

On February 23, 2004, the Pentagon announced that its Defense Criminal Investigative Service would investigate Halliburton to determine whether the Iraq fuel delivery overcharges constituted criminal fraud.[49] The Defense Criminal Investigative Service should investigate the Army Corps of Engineers as well. However, we may never know for sure whether Halliburton's windfalls were due to political cronyism and conspiracy or to multiple levels of profound incompetence. We will never know until someone, somewhere in a position of independent power, starts asking harder questions of the Corps of Engineers staff involved in the oil field contracting for Iraq. American taxpayers deserve more honest answers from the Fort Worth District personnel at the bidders meeting in Dallas on July 14,

2003, and they deserve more honest answers from General Flowers, who is ultimately accountable for the entire U.S. Army Corps of Engineers.

REFERENCES

1. Dan Baum, "Nation Builders for Hire," *The New York Times,* June 22, 2003.
2. Halliburton, "KBR Implements Plan for Extinguishing Oil Well Fires in Iraq," press release, March 24, 2003, posted on the Web at www.halliburton.com/news.
3. "War Funds Advance in Congress," *Engineering News-Record,* April 7, 2003.
4. P.W. Singer, *Corporate Warriors, the Rise of the Privatized Military Industry,* Cornell University Press, 2003; pages 140-147 present an excellent history of the LOGCAP and related contracts.
5. Halliburton, "KBR Reviews Terms and Conditions of Major Contracts in Support of U.S. Troops," press release, January 22, 2004, posted on the Web at www.halliburton.com/news.
6. Shane Harris, "Remaking Iraq," *Government Executive Magazine,* August 15, 2003.
7. Dana Milbank, "For Cheney, Tarnish from Halliburton: Firm's Fall Raises Questions About Vice President's Leadership There," *Washington Post,* July 16, 2002.
8. Jeff Gerth and Don Van Natta Jr., "In Tough Times, a Company Finds Profits in Terror War," *The New York Times,* July 13, 2002.
9. Representative Henry A. Waxman, letter to Les Brownlee, Acting Secretary of the Army, May 29, 2003, posted on the Web at www.house.gov/reform/min.
10. Chip Cummins, "Fight Oil Fires in Iraq Rekindles Desert Rivalry," *Wall Street Journal,* March 31, 2003.

11. Associated Press, "Kuwaiti Team Extinguishes Iraq Oil Fire," *Miami Herald,* April 13, 2003.

12. "U.S. May Drop Plan for Open-Bid Iraqi Oil-Field Contract," *Dow Jones Newswires*, June 4, 2003, cited by Representative Henry A. Waxman in a June 6, 2003, letter to Lt. General Robert B. Flowers, Commander, U.S. Army Corps of Engineers, posted on the Web at www.house.gov/reform/min.

13. Larry Margasak, "Army Backtracks on Halliburton Contract," *Akron Beacon Journal,* June 12, 2003, "Halliburton's Iraq Oil Industry Work May Last Longer Than Expected," *San Francisco Chronicle,* June 13, 2003, and other Associated Press outlets.

14. All U.S. Government contract opportunities valued at over $25,000 are announced on the Federal Business Opportunities Web site (www.fedbizopps.gov). Search on Solicitation Number DACA63-03-R-0021, "Repair and Continuity Operations of Iraq Oil Infrastructure."

15. All of the procurement documents published by the U.S. Army Corps of Engineers' Fort Worth District in association with the competition for the two new Iraq oil field reconstruction contracts (solicitation number and title above) were posted on the Web at https://ebs.swf.usace.army.mil/ebs. The documents are available for free downloading at www.BookRegime.com. The *Restoration of Iraqi Oil Infrastructure Final Work Plan* is included at the end of this book for ease of reference.

16. David Streitfeld, "New Iraq Contracts Offer Just 'Scraps,'" *Los Angeles Times,* August 14, 2003.

17. Representative Henry A. Waxman, letter to Joshua Bolten, Director, Office of Management and Budget, September 30, 2003, posted on the Web at www.house.gov/reform/min.

18. David Ivanovich, "Iraq Contract Award Delayed Again: Decision Probably Slated for Next Year," *Houston Chronicle*, December 2, 2003.

19. Don Van Natta Jr., "High Payments to Halliburton for Fuel in Iraq," *The New York Times,* December 10, 2003.

20. "Halliburton Unit Overcharge May Be $61M," *Dow Jones Newswires,* December 11, 2003.

21. Jackie Spinner, "Army, Halliburton Deny Iraq Fuel Price Allegation," *Washington Post,* October 18, 2003.

22. Congress Daily PM, "No Iraq Supplemental Bills Expected in 2004," *Government Executive Magazine* Daily Briefing, December 17, 2003, posted on the Web at www.GovExec.com.

23. Jackie Spinner, "Halliburton Unit Denies Gouging," *Washington Post,* December 18, 2003.

24. Jackie Spinner, "Senator Says Halliburton's Auditors Saw Problems," *Washington Post,* December 19, 2003.

25. Jim Wolf, "Democrats Press Rumsfeld for Halliburton Records," Reuters, posted on the Web at www.reuters.com.

26. Amelia Gruber, "Calls for Debarring Halliburton Hasty, Analysts Say," *Government Executive Magazine* Daily Briefing, December 22, 2003, posted on the Web at GovExec.com.

27. Jackie Spinner, "Military Ends Halliburton Deal to Supply Gasoline to Iraq," *Washington Post,* December 31, 2003.

28. Neil King Jr., "Army Corps Clears Halliburton in Flap Over Fuel Pricing in Iraq," *The Wall Street Journal,* January 6, 2004.

29. Representative Henry A. Waxman, letter to Lieutenant General Robert B. Flowers, U.S. Army Corps of Engineers, January 6, 2004, posted on the Web at www.house.gov/reform/min.

30. Representative Henry A. Waxman, letter to Condoleezza Rice, Assistant to the President for National Security Affairs, January 15, 2004, posted on the Web at www.house.gov/reform/min.

31. Jackie Spinner, "Corps of Engineers Defends KBR Deal," *Washington Post,* January 13, 2004.

32. Jackie Spinner, "10 Contracts Maintain U.S. Work in Iraq," *Washington Post,* January 15, 2004.

33. Jackie Spinner, "Halliburton Gets More Iraq Work," *Washington Post,* January 17, 2004.

34. Michael Janofsky, "Halliburton Says It Repaid Pentagon $6.3 Million," *The New York Times,* reprinted in the *San Francisco Chronicle,* January 24, 2004.

35. Representative Henry A. Waxman, letter to Representative Tom Davis, Chairman, Committee on Government Reform, January 23, 2004, posted on the Web at www.house.gov/reform /min.

36. Jackie Spinner, "Halliburton to Return $27.4 Million to Government," *Washington Post,* February 4, 2004.

37. U.S. Army Corps of Engineers, "Frequently Asked Questions, Engineer Support to Operation Iraqi Freedom," January 20, 2004, posted on the Web at www.hq.usace.army.mil/CEPA /Iraq.

38. Bruce Stanley, "U.S. In No Rush to Privatize Iraq," *Dayton Daily News,* January 28, 2004, and other Associated Press outlets.

39. Rod Nordland and Michael Hirsh, "The $87 Billion Money Pit," *Newsweek,* November 3, 2003.

40. Jackie Spinner, "U.S. Sets Time Frame for 24 Iraq Contracts," *Washington Post,* November 18, 2003. The official solicitation notices were posted on the CPA's website, www.rebuilding-iraq.net, and on the Federal Business Opportunities website, www.fedbizopps.gov (search for Solicitation HQ0050-04-R-IIRO001).

41. Sherie Winston, "Release of Iraq Rebuilding Contract Delayed," *Engineering News-Record,* December 10, 2003.

42. Neil King Jr. and Yochi Dreazen, "U.S. Postpones $4 Billion in Reconstruction Work," *Iraq Reconstruction Report,* January 8, 2004.

43. Philip Bobbitt, "Seeing the Futures," *The New York Times,* December 8, 2003.

44. Shawn Zeller, "Performance Pay Perils," *Government Executive Magazine,* February 15, 2004.

45. George Cahlink, "Fallen Star," *Government Executive Magazine,* February 15, 2004.

46. Federal Acquisition Council, "History" web page at www.fac.gov.

47. Jason Peckenpaugh, "Focus of New Acquisition Panel Debated," *Government Executive Magazine,* Daily Briefing, February 17, 2004, posted on the Web at GovExec.com.

48. Dwight D. Eisenhower, Military-Industrial Complex Speech, 1961, in *Public Papers of the Presidents, Dwight D. Eisenhower, 1960,* pages 1035-1040.

49. Richard Oppel Jr., "Iraq: Pentagon Opens Criminal Inquiry of Halliburton," *The New York Times*, February 24, 2004.

RESTORATION OF IRAQI OIL INFRASTRUCTURE FINAL WORK PLAN

24 July 2003

Author's Note:

Reproduced here is the Final Work Plan that the U.S. Army Corps of Engineers' Fort Worth District posted for bidders on Friday, August 1, 2003. The unsigned Approval page has been replaced by the signed one posted as a separate document on the same day. The inconsistencies between the Table of Contents and the actual sections of the Final Work Plan, including the lack of Appendices D and E and References A, B, and C, are as they appeared in the document published by the Corps. However, the author believes another document posted by the Corps on August 1, entitled Project Schedule, is the Summary of Costs referred to in the Final Work Plan as Appendix D. It consists of a series of detailed cost spreadsheets too extensive for reproduction here but available for free downloading at www.BookRegime.com. The author also believes that "Table x," missing from the Final Work Plan, refers to one or more of the spreadsheets. Reference C to the Final Work Plan was included in the Corps's Request for Proposal, which can also be downloaded at www.BookRegime.com.

TABLE OF CONTENTS

APPROVAL

Having carefully examined and discussed with our respective representatives the contents of this report and appendices attached therein and here after referred to as "Iraq Oil Infrastructure Restoration Plan" dated on the 24th of July, 2003, we the undersigned acknowledge the plan as follows.

1. The plan is the result of the joint collective effort of the parties listed below:
   - The Iraq Ministry of Oil "MOO" and its associated companies and organizations
   - The U. S. Army Corps of Engineers "USACE" and its Task Force "RIO" staff
   - The Kellogg, Brown and Root staff "KBR"
   - The Iraq Reconstruction and Development Council "IRDC" representatives
   - The Coalition Provisional Authority "CPA" representatives
2. The plan in its present form is considered adequately comprehensive coverage of the total work requirements to restore Iraq oil infrastructure to the pre-war state of the structure and capacity of production, refining and gas processing and oil and gas products distribution.
3. The work requirements are listed in accordance with priorities agreed upon to accomplish production targets in three phases.
4. The method of execution of the work, allocation of responsibility of execution to each party and organizations required to affect execution monitory and control are, as envisioned now, reasonably outlined.
5. The cost estimates of the plan are budgetary and are therefore rough order of magnitude "ROM" estimates. Also, transfer of financial allocations between the various work projects is permissible in view of considering the total work plan as one project.
6. This plan, considered as work initiator, is subject to continuous review and amendments based on impact of events, espe-

cially security aspects, work progress, performance standards of implementation, and financial and funding allocations.

The undersigned therefore approve the plan dated on the 24th of July, 2003, and authorize amendments, agreed upon by our respective parties, that do not conflict with the objectives of the plan. We also pledge to support the restoration effort within the limits of our respective authorities and funding limitations.

Written in English and signed in Baghdad on the day of Thursday, dated 24/7/2003.

Thamir Abbas Ghadhban
Chief Executive Officer (CEO)
Ministry of Oil

Phillip Carroll
Senior Oil Advisor
Coalition Provisional Authority

Robert Crear
Brigadier General, USA
Commanding General
USACE Task Force RIO

EXECUTIVE SUMMARY

The Iraqi Ministry of Oil (MOO), the Coalition Provisional Authority (CPA), USACE Task Force "Restore Iraqi Oil" (RIO) and the USACE Contractor, Kellogg Brown and Root (KBR) conducted a joint planning workshop for the restoration of the Iraqi oil infrastructure on 6-9 July 2003. The meetings were held at the Iraqi Oil Cultural Center in Baghdad. The objective of the workshop was to develop a unified strategic plan for the restoration of the Iraqi oil infrastructure to prewar production capacity. Other objectives were to prioritize, schedule, estimate cost, and assign responsibility for the remaining work required to restore the infrastructure.

In preparation for the workshop and as part of the total planning effort, TF RIO completed development of a 'rough order of magnitude' (ROM) estimate of remaining work which was briefed to the MOO staff on 1 June 2003 and finalized on 3 June 2003 (Reference 3a). The Report stated in Section 3a, "...This report is based on data provided by the Iraqi Ministry of Oil and the detailed assessments conducted by the KBR contractor and TF RIO engineers. It should be noted that this report considered basic infrastructure outside the oil infrastructure (electricity, communications, etc.) to exist and function reliably and not in the scope of the oil infrastructure restoration estimates."

The MOO also developed a plan for the restoration of the infrastructure using the KBR assessments, the ROM, and additional information provided by their operating companies (Reference 3b). The MOO's first step was to assess the damage to each facility using their technical staff. In some cases they relied on specialists or assessments provided by KBR. Their plan concentrates on the upstream sector of the infrastructure as this is the sector that received the most damage, and which is most crucial to restoration of production capacity.

Both plans considered not only actual damage from the war, but also post-war looting, damage and sabotage. We should also note that additional sabotage, damage and looting have occurred since the reports have been completed. Of particular note are the continued and continuing attacks on pipelines and the electrical grid which services production, refining, processing and distribution facilities. The attacks continue to lengthen the time required and increase the cost to restore production capacity.

Participants from the MOO, CPA, RIO, KBR, and the Iraqi Redevelopment Council (IRDC) were assigned to five workgroups as follows:

1. North Group: North Oil Company, North Gas Company and SCOP
2. South Group: South Oil Company, South Gas Company and SCOP
3. Downstream Group: Oil Products Distribution Company, petroleum (Oil) Pipeline Company and Gas Bottling Company
4. Refineries Group: North (Baiji) Refinery, Middle (Daura) Refinery and South (Basra) Refinery
5. Special Companies Group: Iraqi Drilling Company and Oil Exploration Company

Each group met individually in a separate work room, elected a group leader, and, using the two references as a guideline, developed prioritized recommendations on the repairs or equipment necessary to achieve prewar production capacity. Each group then presented its recommendations to a group of Directors and Advisors which made adjustments to ensure that recommendations were within the scope of the mission, to achieve consistency, to approve the project management organization, and to review priorities and amounts. This report summarizes the outcomes of the final, approved recommendations, provides additional analysis and establishes final approved projects, amounts and processes to be followed in executing the work.

The Workshop was a highly successful application of the total infrastructure restoration team. Through the focused application of all team members we were able to develop a comprehensive estimate of the total effort required at all levels of the organization to restore the infrastructure to prewar production capacity.

WORKSHOP SUMMARY

The Iraqi Oil Restoration Planning Workshop was attended by over 100 participants from various disciplines within the Ministry of Oil (MOO), the Coalition Provisional Authority (CPA), USACE Task Force "Restore Iraqi Oil" (RIO) and the US Army Corps of Engineers (USACE) and Kellogg Brown and Root (KBR). This planning group was brought together to focus on identifying and prioritizing oilfield infrastructure requirements that would result in immediate and short-term Crude Oil production for both internal and external needs.

The group was broken into six (6) separate disciplines, to include, South Oil Company, North Oil Company, Northern Gas Company, Refineries, Drilling and Exploration.

For a period of four (4) days, the various groups made up of experts and professionals within the specified disciplines, discussed, reviewed, analyzed and prioritized restoration programs and projects focusing on re-establishing the Oilfield Infrastructure capabilities to pre-war levels of crude oil production and refined products.

This list was then prioritized once again and assigned a time schedule and/or phase of execution based on the immediacy that the work item will contribute to the restoration of production or refining capacity. The Phase 1 projects therefore are those identified projects that will restore the infrastructure capabilities to pre-war levels in the shortest period of time.

As a result of this combined effort, a total of 220 projects and/or items to be procured, were identified for a total estimated cost of $1.144 billion. See Table 1 for details.

North Oil Company (NOC) and South Oil Company (SOC) projected expenditures that totaled $652 Million or 57% of the total budget. Of this total, there was a contingency project/budget identified and prioritized by the South Oil Company in the amount of $193.5 million USD. See Table 2 for details. (Note: Contingency costs have not been included in any of the cost calculations or spreadsheets.)

TABLE 1

Company	No. of Projects	Estimated Cost (Million US$)			
		Phase 1	Phase 2	Phase 3	Total
North Oil Company	48	234.250	61.500	24.500	320.250
South Oil Company	57	222.679	73.000	36.200	331.879
South Gas Company	9	23.750	0	0	23.750
North Gas Company	7	25.400	6.500	0	31.900
Refineries	29	22.748	9.259	10.940	42.947
Downstream	29	102.320	34.267	12.266	148.853
SCOP	4	41.500	2.000	0	43.500
Iraq Drilling Company	26	15.360	24.150	92.500	132.010
Oil Exploration Co.	11	28.226	40.250	0	68.476
Total	**220**	**716.233**	**250.926**	**176.406**	**1143.565**

TABLE 2

Contingency (Million US$)		
South Oil Company	Other Pipeline Work	7.000
South Oil Company	Temporary Power	96.500
South Oil Company	Contingency	90.000
Total		**193.500**

The total expenditures projected for Communications was $45.9 million USD. Communications were identified and prioritized by the Downstream Operating Companies, South Oil Company and North Oil Company. Please see the Table 3 for details.

TABLE 3

Communications (Million US$)			
Company	**KBR**	**Iraq Oil Ministry**	**Total**
Oil Pipeline Company	0.250		0.250
Gas Filling Company	0.130		0.130
South Gas Company	0.500		0.500
South Oil Company	21.000	9.000	30.000
North Oil Company	10.000	2.000	12.000
North Oil Company	2.500	0.500	3.000
Total	**34.380**	**11.500**	**45.880**

In summary, expenditures were categorized into 2 separate groups; projects category (1) and procurement category (2). These two categories combined, make up the whole budget of $1.292 billion USD. Procurement is further divided into items to be procured by KBR and/or the Ministry of Oil (MOO). The commodities identified for procurement by KBR total $631 million or 55% of the total budget. The project construction identified for procurement by KBR total $320 million USD or 27% of the total budget. See Table 4 for details.

TABLE 4

Commodity vs. Project Costs (Million US$)				
Company	**KBR Commodity**	**KBR Project Procurement**	**MOO EPC & Local Procurement**	**Total**
North Oil Company	175.800	99.425	43.025	320.250
South Oil Company	46.700	175.700	109.479	331.879
South Gas Company	9.750	8.400	5.600	23.75
North Gas Company	30.900	0.700	0.300	31.900
Refineries	28.032	77.240	7.675	42.947
Downstream	98.710	28.470	21.673	148.853
SCOP	43.500	0	0	43.500
Iraq Drilling Company	132.010	0	0	132.010
Oil Exploration Co.	64.950	0	3.526	68.476
Total	**630.352**	**319.935**	**193.278**	**1143.565**

(Note: For MOO EPC & Local Procurement, most of the estimate will be funded through the MOO budget.)

Expenditures budgeted for transportation to include vehicles, heavy equipment, cranes, buses, tank trucks and firefighting equipment are provided below [Table 5].

TABLE 5

Transportation (Million US$)				
Company	**Heavy Equipment & Vehicles**	**Firefighting Equipment**	**Forklifts & Tankers**	**Total**
North Oil Company	85.000	15.000	0	100.000
South Oil Company	19.000	0	0	19.000
North Gas Company	15.000	0	0	15.000
Refineries	0	0	0.183	0.183
Downstream	5.000	0	7.500	12.500
SCOP	20.000	0	0	20.000
Iraq Drilling Company	5.000	0	0	5.000
Oil Exploration Co.	0	0	0	0
Total	**155.000**	**15.000**	**7.683**	**171.683**

A total expenditure of $171.6 million USD has been estimated to be the cost of mission essential transportation and heavy equipment (pickups, Sport Utility Vehicles (SUV), busses, cranes, loaders, forklifts, etc.) necessary for sustained production. The North Oil Company has projected a total requirement of $100 million or 58% of the total $171.6 million projected for transportation. The total $171.6 million projected for transportation relates to 14% of the total $1.242 billion budget.

(Note: Review of equipment contracted and scheduled for delivery, through the Oil for Food program may eliminate the need for some of mentioned equipment, i.e., Firefighting Equipment.)

PROJECT EXECUTION PLAN

OVERVIEW

During the MOO/RIO Team Workshop held 6-9 July 2003, it was apparent that the MOO operating companies and SCOP, the MOO engineering and construction firm, elected to be responsible for the engineering and construction of the largest majority of the USACE funded projects. The operating companies within the Ministry of Oil will be the client with overall responsibility while KBR will provide primarily a procurement role for restoration commodities. Either the operating company or SCOP will serve as the EPC agent for most projects. KBR will serve in a consulting role for some projects under the purview of SCOP and the operating companies, providing expertise as appropriate, and will be the EPC agent for a small number of selected projects.

These expectations provide a series of challenges for USACE funding and management of the restoration process within the guidelines for contracting, expenditure control, and scheduling responsibilities expected by the US Department of Defense, CPA, USACE, and MOO. It was also clear that the MOO objective dates for the completion of the projects (Phase 1 by 30 Sep; Phase 2 by 31 Dec; Phase 3 by 31 March) will be a challenge to the entire team. Senior management of MOO stated that the final decision as to which company will be responsible for execution of each project will be determined by the Ministry based on technical and staff capabilities, preference of the operating company, other workload, scheduling, and cost effectiveness, as they deem most appropriate.

In addition, cash management will be very important in the early phases of project execution. The funds for this work are tied to the US appropriations process so it will be necessary to insure that funds available between July and 30 Sep 2003 are scheduled for the highest priority Phase 1 projects first. Phase 2 and Phase 3 projects can be scheduled as additional funds become available.

For the several parties involved in the oil infrastructure restoration process, there are various considerations that may impact the expectations and concerns of others. In order to address who will do what and how it will be done, the primary objectives are:

- Assure timely completion of project requirements for the operating companies
- Provide most cost effective execution of work on a minimal cost plus basis
- Give MOO proper oversight of life cycle facility performance and quality concerns
- Assist SCOP to rebuild the operating companies and its technical and project control capabilities through phased technology transfer, hardware and software transfer, and on the job training
- Assure USACE and MOO of timely execution, and acceptable standards of management, cost control, quality assurance, environmental, and safety

PROJECT SCOPE

There were two types of projects identified during the Workshop for the various companies. The first type was "procurement only" and second type was "EPC facilities" (Engineer-Procure-Construct). In Table x, the EPC projects are identified by a "1" in the category column and the procure projects are identified by a "2."

The greatest majority of the funding needs for these projects include purchasing of various commodities for the operating companies and equipment and material needs within the scope of the EPC projects. The commodities for the operating companies include looted and damaged vehicles and buses, transport and heavy equipment, fire and safety gear, spare parts, bulk mechanical and electrical material and equipment, and stand-by power generation. In the EPC work, there will be bulk materials such as pipe, valves, and fittings in addition to equipment such as vessels, pumps, compressors, motors, and control systems.

EXPECTATIONS AND RESPONSIBILITIES

Ministry of Oil

MOO expects that CPA and USACE will assist in providing the resources to restore the Iraqi oil infrastructure to the levels of production that will allow suitable exports for revenue and meet the internal needs of the country of Iraq. The Chief Executive Officer of the MOO has asserted that MOO is the final authority for the planning and execution of the restoration projects in Iraq. In this understanding, MOO has made it clear that final agreement of scope and engineering of facility restoration work will be made within the Ministry rather than individual operating companies. For this reason, it is the desire of MOO to assure that SCOP provides the engineering and construction management of all major projects with assistance from KBR only as requested. It has been recognized by MOO that KBR is the logical entity to perform the procurement function with joint technical approval by the operating company or SCOP as the situation requires.

Operating Companies

The operating companies include SOC, SGC, NOC, NGC, and the pipeline and distribution companies. It is their objective to return to full capacity service in the shortest time possible. They have helped KBR with facility assessments in the field and have provided their operational expertise to help provide temporary solutions and "band-aid fixes" for damaged and looted facilities. Where facilities must be repaired or rehabilitated while maintaining operations, it would probably be most prudent to allow these operators to maintain full responsibility for reconstruction or repair within these boundary limits. In such cases, SCOP may be able to provide certain technical assistance such as QA/QC inspection while the operating company supervises the actual site-work and subcontractors.

SCOP (State Company for Oil Projects) Involvement

As the engineering and construction agency for the Ministry of Oil, SCOP serves the operating companies for planning technical facility needs, providing engineering expertise in major disciplines, performing project management and project control functions for project execution, detailed design and construction management and construction/ commissioning of project facility implementations. During the period of economic sanctions, SCOP served as the prime contractor for most of the new project development and expansions in the Iraq oil infrastructure. The sanctions have limited SCOP access to some information technology, computer aided engineering, and project management techniques. Likewise, database information involving vendors, equipment, and general technical data has not been available to SCOP. It is desirable to equip SCOP with modern computer and information assets and to make them consistent with KBR project management tools to promote information sharing on those projects where KBR and SCOP will be working in close cooperation.

USACE Issues

The US Army Corps of Engineers has been designated as the responsible Department of Defense agency to plan and execute the restoration of the Iraq oil infrastructure to pre- war levels of production. USACE must consider the legal and procedural regulations for handling Congressional appropriations for conducting RIO operations in Iraq. The expenditure of appropriated Department of Defense funds will need to meet the audit requirements for the USACE and its contractors.

In addition to various funding issues, there are other engineering requirements that must be addressed under the responsibility of the USACE. These obligations typically address the management and accountability of project needs to assure mission success, environmental and safety standards for the project and its personnel, quality standards and professional engineering practices, and other contracting and engineering guidelines under law, regulation, ethics, due diligence, and sound engineering judgment.

RIO must also give precedence to timely completion of quality facilities, not only because delays or changes may impact available funding resources, but also because timely completion has secondary impact on the speed of economic recovery, security, and the quality of life of the Iraqi people. All of these bear directly on the success of the mission of the Coalition Forces.

Kellogg Brown & Root (KBR) Commitment

As the Government Operations Contractor selected by USACE for the RIO project, KBR must satisfy all legal and regulatory requirements in the execution of the oil infrastructure restoration work. The KBR procurement and subcontracting procedures developed over several years of Government contracting service have been approved and have been verified during several annual audits. Support of the USACE commitments to the Iraq MOO can be made via the KBR RIO government contract. The procedures to provide the funding support must meet the stipulations of the RIO contract and other procurement and contracting regulations governing this type of contract and support. To conform to these procedures, it will be necessary for MOO to assure their proper documentation and internal procedures for tracking and controlling costs of funds and procurement items provided to them under the KBR contract. Contracting arrangements will therefore need to reflect the appropriate guidelines and procedures for support, procurement, engineering, and construction.

TEAM APPROACH TO MOO, SCOP, AND KBR OPERATIONS

RIO has given KBR the goal to maximize the use of Iraqi manpower, supervision, technical know-how, and private or public company involvement in the execution of the restoration plan. To this end, KBR will work in alliance with MOO, SCOP, and the other operating companies. This includes embedding SCOP or other MOO personnel in KBR engineering, project control, procurement, and construction supervisory teams or functions in order to phase-in Iraqi personnel to use of approved project management and cost control systems and technical procedures. As SCOP transitions to use these approved computer/communications and other information technology processes, they will assume progressively greater project management responsibilities. This training and support program may include exchange of personnel, survey of management procedures, database review and updates, etc. On a milestone basis the program would be intended to transfer to SCOP the equipment and process common to all international oilfield contractors and acceptable to institutions such as the World Bank, EX-IM Bank, the United Nations Development Fund, and others.

The secondary goals of this "technology transfer" process will be to ensure that USACE is assured of mission success through the use of management tools it has approved; KBR would have a better opportunity to provide satisfactory service to both MOO and USACE; and that MOO and operating companies can manage the remaining projects using similar tools and processes with which they were managed from the outset.

PROCUREMENT DEFINITION & EXECUTION

The operating company assigned by MOO for overall responsibility of the "procurement only" type projects will have responsibility for final definition of quantities and specifications of the various commodities. The total commodities cost must include the purchasing management, freight traffic handling, and customs clearances. All such associated costs will need to be within each project budget.

It will be most prudent for KBR to manage their primary planning and initial procurement work from Kuwait City due to communications, travel, and security arrangements currently in place. Later as freedom of movement and telecommunications can support transitioning of procurement activities to a more central control point in Baghdad, a transition of these operations can be planned and implemented with a goal of minimizing interruptions to the procurement operations. KBR Area Coordinators will need to work directly with Iraqi companies in primary areas of operations and headquarters locations. These should be aligned with the Project Offices in the North and South of Iraq and the Central area of Baghdad to include all company requirements directed from the Ministry of Oil. These areas and company responsibilities would include the following:

North Project Office	South Project Office	Central Project Office
North Oil Company	South Oil Company	Ministry of Oil
North Gas Company	South Gas Company	SCOP
Baiji Refinery	Basra Refinery	Oil Pipeline Co.
		Oil Prod Distribution Co.
		Gas Filling Co.
		Daura Refinery
		Iraq Drilling
		Iraq Exploration
		Other Companies

It is recognized that an extensive amount of purchasing coordination will be required between KBR and the operating companies over a relatively short time period. Procedures will need to be developed between KBR, SCOP, and the operating companies to assure suitable control and scheduling in an efficient and effective manner from consolidating common items for purchase to completing authorized requisition requests and issuing approved purchasing orders and shipping instructions. Good communications and close cooperation will be essential.

Area Purchasing Coordinators will report to the KBR RIO Procurement Manager. The initial office for this key role will be in Kuwait where current operations are already in process. Later, this role will be transitioned to the Central Baghdad office as freedom of movement, vendor visits, and suitable telecommunications can be deemed ready for the transition of this function to the Central office. It is recommended that an Area Coordinator would be appointed to each North and South Project Office and to the Baghdad Central Office. Senior Buyers may be appointed to assist these Area Coordinators in the initial planning and scoping of the purchasing needs before returning to other Buying Offices in Kuwait, Turkey, Houston, Dubai, or elsewhere. Clerical and other support staff, including RIO Project Engineering, Project Controls, and Operations will provide general and technical support as needed to meet the needs of the overall procurement functions.

An outline of the KBR procurement process is at Appendix C. The basic staff would therefore include the following:

- KBR RIO Procurement Manager – Main Program Office with travel to Area and Buyer offices
- North Area RIO Purchasing Coordinator – Kirkuk with travel to Kuwait and Turkey
- South Area RIO Purchasing Coordinator – Kuwait with travel to Basra (unless camp facilities allow for Basra billeting with occasional travel to Kuwait)
- Central RIO Purchasing Coordinator (Deputy Procurement Manager – Baghdad with travel to Kuwait, Turkey, Houston, Dubai, as required
- Senior Buyers & Expeditors – Home or Field Office Assignments
- Clerical Support – Kuwait/Kirkuk/Basra/ Baghdad, as required

EPC PROJECT EXECUTION

There will be immediate collaboration with key SCOP and operating company engineers on all work currently underway by KBR. KBR will immediately involve SCOP and operating company engineers into the various project teams now underway in KBR. This will assist in better understanding of MOO, SCOP and operating company practices and standards of operation and design preferences and a better understanding by the Iraqis of the project particulars dictating design and schedule. Key Iraqi engineers will participate within these design teams. This will give an added benefit in seeing work progress with proven project management procedures, using the software, management systems, information systems, and computer aided design packages used by KBR.

While the critical restoration work is underway, KBR will assist, as requested, with on-the-job training for current project management and procurement functions. This would include updating or development of procedures for Project Management Control and Accountability, Time Keeping and Cost Control, Project Controls and Planning/Estimating, QA/QC, Document Control, Information Technology, Web based internet applications, and others. An ISO 9001 type audit of SCOP practices and procedures could be part of this survey and proposed process review.

In addition to the management assistance, the KBR IT staff will complete acquisition of computers, printers, plotters, scanners, faxes, and state of the art software for installation and training of SCOP staff and technical management personnel. Technical assistance for CAD drafting and engineering software may be included in this scope along with instruction for modeling and simulating system facility designs. Arrangements will also be made to work with MOO management on designing a Local Area Network and a technology needs assessment study for software and hardware requirements.

PROGRAM MANAGEMENT PLAN

PURPOSE

The restoration of the Iraqi oil infrastructure will be executed as a team effort. The team will include MOO, MOO subsidiary operating and engineering & construction companies, OCPA, USACE-RIO, and the US Government Services Contractor, KBR.

PROJECT MANAGEMENT TEAMS

Figure 1, which follows, includes an organization chart for the proposed Program Execution Plan. There will be seven project groups working to oversee the execution of the many projects to restore the Iraqi oil infrastructure in each of the operating companies. Three of these groups will be supporting operations of the operating companies on a regional basis. These will include the North Office in Kirkuk, the South Office in Basrah, and the Central Office in Baghdad. Three other groups will be of a functional nature providing specialized support to the Project Office Teams. These functional teams include Procurement, Project Controls, and Technical Support. The guidance and direction of these teams will be provided by the Program Management Team, consisting of MOO, USACE, CPA, SCOP, and KBR key project management personnel.

The make-up of each team and its primary work area will include team member personnel as shown in each block. The responsibilities are also summarized by function and operating company for the required restoration work. See Figure 1 for an organization chart.

FIGURE 1
PROGRAM EXECUTION PLAN

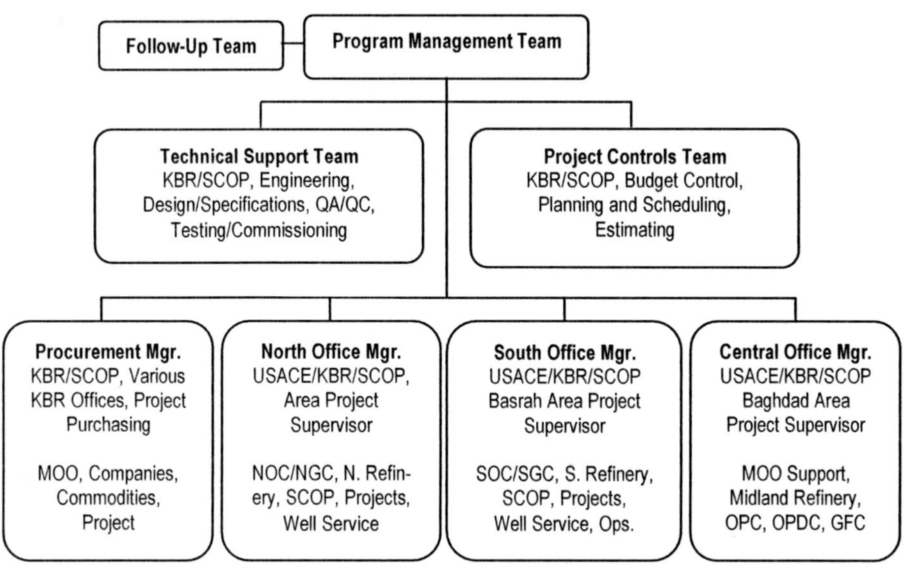

FOLLOW-UP PLAN

Measuring the success of the Project Execution Team's efforts will include different types of benchmarks. These may include achieving targets for:

- Oil production levels
- Oil and product export volumes
- Milestone schedule results
- acceptable quality standard performance
- Satisfactory safety and security results
- Environmentally acceptable performance
- Effective cost monitoring and results
- Timing to eliminate product imports
- Reliability of continuous stable operations
- Timely suitable services to industrial and power plant customers
- Team member and public satisfaction with overall team performance

A procedure for measuring the success of the Project RIO team effort will be part of the responsibility of the "Follow-Up Team."

The organization of the team is displayed in Figure 2.

FIGURE 2
FOLLOW-UP ORGANIZATION

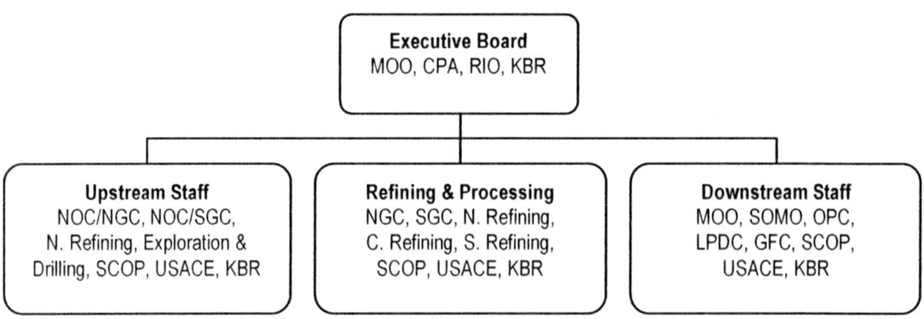

FOLLOW-UP TEAM

(a) Responsibilities

The Operating Staff will be responsible for assessing results and troubleshooting bottlenecks or restraints to the execution of technical and operations work involved in the restoration of the Iraq oil infrastructure.

- Serve as advisors to the Executive Group
- Resolve project bottlenecks and disputes
- Adjust schedules within main project/phase durations
- Cost/budget adjustments up to a set limit
- Hold project activities due to noncompliance with technical standards
- Recommend changes when activities are detrimental to sustained production levels
- Recommend personnel changes when appropriate to improve team performance
- Coordinate commissioning and start-up/shut-down planning to minimize interference with adjacent operations
- Validation of progress and status/completion reports

(b) Follow-Up Team Provisions

Time and resources will be provided for each member of the Project Management Team to perform their duties. This will include cooperation in accommodating priority schedules for meetings, work locations, communications, air transport, vehicles, software, and hardware.

Each team member will be included on the reporting distribution of the area of operations within each member's area of responsibility. This will include project master plans and budgets, weekly and monthly reports, regular situation reports, and completion reports and schedule updates.

Regular site visits will be arranged for individual and team reviews. Conference calls and regular verbal reports will be provided as well as any third party QA/QC or HSE activity reports. Each team member will be provided with email access and computer software to allow for access to reports and information.

A regular written report will be made to the Executive Team summarizing progress and target results with a formal report every month. It is intended that the report will address the overall status of the infrastructure restoration for each operating MOO Operating Company with a summary of budget and mission successes.

CONTRACTING ARRANGEMENTS

An outline of the KBR contracting process is at Appendix D. It will be modified to provide for MOO, SCOP and Operating Companies to be involved in the contracting process as appropriate. Specifically, at a minimum, there will be direct Iraqi input at two points in the process: (1) the appropriate company will either provide specifications or will approve product specifications prepared by KBR. At the same time they will provide a list of preferred vendors, if available. (2) KBR will provide the bid analysis to the appropriate company for approval prior to final decision. In other instances KBR and the appropriate company may have personnel working together throughout the procurement process. The contracting arrangements will provide for maximum utilization of the Iraq Ministry of Oil, their operating companies, SCOP, and other public or private companies in Iraq to employ primarily Iraqi labor within the limits permitted by legal and regulatory guidance.

During the Workshop, the USACE Project Offices and the MOO operating companies provided lists of commodities required for restoration of operations and projects needing engineering, procurement, and construction to replace damaged facilities and operations. KBR and the USACE Project Offices worked with these operating companies to identify proposed solutions and provide estimates of cost and schedule on the major projects identified. Various contracting alternatives were considered to propose a means to meet the expectations of MOO. Types of contracting arrangements included:

- Cost Reimbursable
- Fixed Price

- Unit Price
- Rates Basis of Charge

The proposed methods of execution, using KBR as the contracting authority for USACE, could include the following:

Purchasing

The execution of the procurement activities identified as the responsibility of USACE's contractor are fairly straight forward and will be handled in the following manner:

USACE will task its contractor, to coordinate with the MOO, SCOP and its operating companies regarding the specifications and technical aspects of the procured items. MOO and its operating companies will be responsible for supplying complete technical specifications and suggested vendors in a prompt manner allowing the USACE contractor to fulfill the request and maintain the integrity of its procurement process according to its approved procedures while maintaining the schedule for the delivery of the items. KBR will also request the appropriate operating company's approval of the recommended bid analysis prior to final purchase.

Engineer-Procure-Construct (EPC) Type Projects

EPC projects will be assigned to either SCOP, an MOO Operating Company or KBR. In those instances where it is an MOO organization that is responsible, most costs will be funded from the MOO operating budget and the responsible organization will be required to develop a project execution schedule and a spending schedule. KBR will assist these organizations with project scheduling, resourcing and spending plans as requested. KBR will also assist with the procurement of equipment or materials outside the scope of the operating project that are necessary to project implementation.

APPENDIX A

IRAQI OIL INFRASTRUCTURE RESTORATION PLAN WORKSHOP AGENDA

Date	Time	Activity
5 JUL		Travel to Baghdad
6 JUL		Travel to Oil Cultural Center
	0830	Introductory Remarks: BG Crear, Dr. Thamir Al-Ghadhban and Gary Vogler
	0900	Workshop methodology and objectives: Gary Loew
	0945	Work group and work area assignments: Bob Corrales i. No. Group: North Oil Co., North Gas Co., SCOP ii. So. Group: South Oil Co., South Gas Co., SCOP iii. Refineries Group: North, Middle and South Refining Companies iv. Downstream Group: Oil Products Distribution Co., Oil Pipelines Co., SCOP, Gas Bottling Co. v. Specialized Companies Group: Iraqi Drilling Co., Exploration Co.
	1100	Communications Brief by KBR
	1130	Lunch
	1300	Group Work Sessions
	1600	Daily Issues and Wrap-up Session: Directors, Advisors, Group Leaders and Facilitator – Nabil Lammuza
7 JUL	0900	Group Work Sessions
	1130	Lunch
	1230	Specialized Companies Group presentation to Directors and Advisors: Iraqi Drilling Co., Exploration Co. (Other work sessions continue as needed)
	1600	Daily Issues and Wrap-up Session: Directors, Advisors, Group Leaders, Facilitator – Gary Loew
8 JUL	0900	Downstream Group presentation to directors and advisors: Oil Products Distribution Co., Oil Pipelines Co., SCOP, Gas Bottling Co. (Other work sessions continue as needed)
	1130	Lunch
	1230	Refineries Group presentation to directors and

		advisors: North, Midland and South Refining Companies (Other work sessions continue as needed)
	1600	Daily Issues and Wrap-up Session: Directors, Advisors, Group Leaders, Facilitator — Nabil Lammuza
9 JUL	0900	South Group presentation to directors and advisors: South Oil Co., South Gas Co., SCOP (Other work sessions continue as needed)
	1130	Lunch
	1230	North Group presentation to directors and advisors: North Oil Co., North Gas Co., SCOP
	1500	Follow-Up Team Charter and Assignment
	1600	Closing Remarks: Gary Vogler, BG Crear, Dr. Thamir Al-Ghadhban

Team	Ministry of Oil	CPA Oil	TM RIO	KBR
Directors	Dr. Thamir Al-Ghadhban	Gary Vogler	BG Robert Crear	Stoney Cox
Advisors	Nabil Lammuza Ahmad Al-Shamaa Hazim Sultan Dr. Salim M. Juma Sabah Juahar Hussain Fatal Saad Abdulwahab	John Kjar Dave Callahan MAJ Freeze	Gary Loew COL Jenkins Don Dunn Tom Armstrong Stan Reese Ralph Schuler	Doug Fletcher **IRDC** Muhammet Ali-Zainy Hama-Jaza Kader
North— NOC, NGC, SCOP	NOC 4-8 Reps NGC 3 Reps SCOP 6-7 Reps		Doug Cox MAJ Joe Hanus Mike Demasi	Wes Jordan Moaid Salum
South— SOC, SGC, SCOP	SOC 4-8 Reps SGC 3 Reps SCOP 6-7 Reps		Johnny Rozsphal Lori Thomas MAJ Langlois	Ed Johnson Samy Ghattas
Downstream— LPB Bottling Co., Oil Products Distribution Co., Oil Pipeline Co.	+2 Aziz Al Ghanmi + 2 +3		Ron Brock Jim Gilmore	John Stout Doyle Sanders
Refineries— Baiji Dora Basra	Basim Al Furati Amira M. Jawad Dr. Ali M. Abbas Abdul Jabar Al-Shekh Dhia Qumar Mahdi Al-Qarhhulia 2-3 Reps No. Ref. 2-3 Reps Midl. Ref. 2-3 Reps So. Ref.		Scott Ross McKenzie	Bill Tyson
Special Companies— Oil Exploration Co., Iraq Drilling Co., Reservoirs & Fields Development	Dr. Sakim Moo +2 +7		Jay Leighton	Kevin Wilson Roger Knight Joseph Ziegler Paul Porter

APPENDIX B

IRAQI OIL INFRASTRUCTURE RESTORATION PLANNING WORKSHOP
MINISTRY OF OIL — OIL CULTURE CENTER
BAGHDAD — IRAQ
ATTENDANCE LIST JULY 6–9, 2003

MINISTRY OF OIL/STUDIES, PLANNING & FOLLOW-UP DEPARTMENT

1. ABBAS, NADHIM
2. AL- OBAIDI, ALI
3. AL-ANI, RIADH
4. AL-GHANIMI, ABDUL-AZIZ
5. AL-JAWHER, SABAH
6. AL-SHAMMA' AHMAD
7. EDWARD, LINA
8. FADHIL, ABEER
9. ISTEFAN, AMJAD
10. JUM'AH, SALIM
11. LAMOZA, NABEEL
12. MADLOOL, HAMID
13. MAJEED, SHAHZINAN
14. JAMEEL, SHAHNAZ

MINISTRY OF OIL/TECHNICAL DEPARTMENT

1. ABDUL-WAHAB, SA'AD
2. AL-FURATI, BASIM
3. AL-HILFI, MAJEED
4. AL-KHAYAT, FALIH
5. AL-SHBENDER, ALI
6. AL-TA'I NAJIM
7. AQRAWI, AKRAM
8. JALEEL, MANSOUR
9. SAFER, DHIA'
10. YA'QOOB, RAJA'

MINISTRY OF OIL/NATIONAL INDUSTRY DEPARTMENT

1. AL-ANBAKI, QAHTAN
2. AL-SHAIKHLI, A.

3. JABBAR, A.
4. MOSA, NIHAD

MINISTRY OF OIL/ADMINISTRATIVE DPARTMENT

1. ABDUL –QADER, SABAH
2. AL-SAMARA'I, BAHA'
3. KHALAF, MAHMOUD
4. MARIOSH, SATTAR
5. MUBARAK, AMIR

MINISTRY OF OIL/EXECUTIVE DIRECTOR OFFICE

1. SULTAN, HAZIM

MINISTRY OF OIL/RESERVOIR DEVELOPMENT DEPARTMENT

1. ABDUL LAH DALMAN
2. AL-FATTAL HUSSAIN

SOUTH OIL COMPANY

1. ABBAS, SABAH
2. AL-QURAISHI, ODAI
3. HUSSAIN MAJEED
4. RAMADHN, FAIQ
5. SALIM, YAS

NORTH OIL COMPANY

1. AL- BAGHDADI, HASHIM
2. AL- DABBAGH, HISHAM
3. AL-SA'EDI, HAMEED
4. JAWAD, WALEED
5. JOMA'AH, NAWZAD
6. KHALID, MOHAMMED
7. RAZZAQ, HAITHAM
8. SALMAN, QASSIM

SOUTH GAS COMPANY

1. BARSACH, AWANES
2. IBRAHEEM, ABDUL-A'OF
3. ZUCHAIR, ABDUL-LATEEF

NORTH GAS COMPANY

1. ABDUL-WAHAB, KHALID
2. HASSAN, HAWNER
3. NAJEEB, MUNEER

OIL PROJECTS COMPANY

1. AL- BADRI, MUTHANNA
2. ALWASH, ALI
3. BAGHDADI, M.
4. HASSAN, A.
5. HAWA, NAHIDH
6. IHSAN, KAMERAN
7. KAMEL, NIDA
8. KASHMOLAH, HUSSAM
9. KHALAF, NATIQ
10. SA'AD, ALI
11. YALDA, SALIM

MIDLAND REFINERIES COMPANY (DORAH)

1. NORI, SA'AD
2. SA'EED, SALAH
3. MUHSIN, SALAH
4. MOHAMMAD, JA'FAR

SOUTH REFINERIES COMPANY

1. SALIM, JAMAL
2. MA'TOOQ, MAHDI

NORTH REFINERIES COMPANY

1. AHMAD, MOHAMMAD
2. AMIAZON, NADEEM
3. RASHAD, SAMI

IRAQI DRILLING COMPANY

1. HANDOOL, TALIB
2. SHAKER, ZUHAIR
3. WAFI, ALA'

GAS FILLING COMPANY

1. ABBOOD, ABDUL-KAREEM

2. AL-UR'IB, SALAH
3. HABEEB, HABEEB

OIL EXPLORATION COMPANY

1. AL-MUKHTTAR, ALA'
2. MUSTAFA, BASSIM
3. NORI, SABAH
4. RIZOQI, SALIM

OIL PRODUCTS DISTRIBUTION COMPANY

1. FLAIEH, SA'AD

OIL PIPELINE COMPANY

2. ALI, HUSSAIN
3. AL-SHUJAIRI, IBADI

KBR [HALLIBURTON KBR]

1. BRUSS, LARRY
2. COX, STONEY
3. FLETCHER, DOUGLAS
4. GHATTAS, SAMI
5. JORDAN, WES
6. JOHNSON, ED
7. PORTIN, PAUL
8. SALUM, MOAID
9. SANDERS, DOYLE
10. STOUT, JOHN
11. TIDER, JOSEPH TYSON, BILL
12. WILSON, KEVIN , Halliburton

CPA — OIL

1. ABDUL-MUHSSIN, DHAHIR
2. KJAR, JOHN
3. VOGLER, GARY

IRDC — CPA

1. ALEXANDER, PAKEZA
2. KADIR, HANNA
3. ZAINI, MOHAMMAD

CORPS OF ENGINEERS, TEAM RIO [RESTORE IRAQI OIL]

1. COX, DOUG, Northern Area Office
2. HERRING, LEWIS, Northern Area Office
3. HANUS, JOE, MAJ, Northern Area Office
4. DeMASI, Mike, Southern Area Office
5. ROZSYPAL, JOHNNY , Southern Area Office
6. CREAR, ROBERT, BG, RIO, Main
7. DAVIS, IAN, LT, RIO Main
8. LOEW, GARY, Rio Main
9. SCHULER, RALPH, RIO Main
10. ARMSTONG, THOMAS, RIO Main
11. DUNN, DON, RIO Main
12. GILMORE, JIM, RIO Main
13. MECKENZIE, JIM, RIO Main
14. BROCK, RONN, RIO Main
15. MILES, CHARLES, RIO Main
16. REESE, STAN, RIO Main
17. ROSS, SCOTT, RIO Main
18. SPAIN, KENNY, RIO Main
19. THOMAS, LORI, RIO Main
20. WRIGHT, STEVE, RIO Main
21. CORRALES, ROBERT, CPT, Rio Main

CENTCOM

1. LEIGHTON, JAY, MAJ.

APPENDIX C

THE KBR PROCUREMENT PROCESS

Material Requisition

Required for all procurement actions. Must be prepared by the Requisitioner and approved by someone with the authority to approve material requisition, BEFORE any procurement action may be initiated. Must be completely filled out and include the following:

1. Is there a detailed description of the items required/scope of work?
2. Has the need been approved by someone with Requisition authority?
3. Is there a required delivery date?
4. Is there a budgetary estimate?
5. Is there an independent estimate, if single source?

Single Source Justification

It is KBR's practice to maximize full and open competition. Single source procurements are an exception to both KBR practices and government regulations.

Responsibilities

Requisitioner – Responsible for providing procurement a written justification that clearly defines the acceptable reasons for only going to one source. This memorandum must be approved by the Project Manager.

Buyer/Subcontractor Administrator – Responsible for ensuring that the single source justification is rational, the approach is logical given the circumstances or that the goods/services cannot be procured from any other source (sole source).

Reasons for Single Sourcing – The following are acceptable reasons for single sourcing:

1. The item required can only be procured from one source

2. Unusual and compelling urgency
3. Continuation of an existing service or compatibility
4. Supplies or services which are authorized or required by statute to be bought from a specific source

Single/Sole sourcing is the exception – not the rule. There must be an executed Sole/Single Source Justification in the file.

Request for Proposals/Quotes

1. Once material requisition is received – Procurement issues an RFP to obtain pricing for both new work and change orders;
2. Procurement is the SOLE communicator with bidders during RFP stage
3. For new work, Procurement must insure adequate competition
4. For change orders, Procurement must have detailed scope of work for change order and independent estimate to justify the price
5. Check to make sure bidders are not on Federal Debar listing prior to sending RFP (http://eplis.arnet.gov/)

IMPORTANT: All bidders must be given the same opportunity of receiving award. All bidders should receive the same information in order to preclude one bidder from having an advantage over the other bidders.

IMPORTANT: It is KBR policy that information concerning subcontractor/suppliers offers and prices paid shall be treated as confidential. NO DISCLOSURE OF ANY PRICING.

Evaluation of Bids/Proposals

1. Determine responsiveness of proposal – check to ensure all required submittals are included – performed by Procurement and Operations.
2. Perform Price/Cost Analysis performed by Procurement
3. Have technical evaluation performed by technical staff, if necessary

Price Analysis

It is KBR practice and government regulations mandate that all purchases/ sub-contracts be awarded at a fair and reasonable price.

Every procurement action must have some form of price analysis performed in order to determine if the price is fair and reasonable.

Methods used to determine if a price and reasonable:

1. Adequate competition
2. Comparison to previous competitive price paid
3. Comparison to prices of similar items
4. Catalog or market prices
5. Prices set by law regulations
6. Comparison with a valid independent estimate

Two or more prices do not guarantee reasonableness of price.

The following are some instances where you may not have adequate competition:

1. Price variance among proposals is in excess of 25%
2. When the lowest price received is unreasonably low
3. When all prices received are considered too high
4. When one firm has a competitive advantage (i.e., local firm vs. out of town contractor)
5. Evaluation should ensure that identical FOB Terms were bid by all offerors (i.e. freight included/excluded with same FOB Point)

Negotiations

1. May be required in order to determine fail price for single sources and change orders
2. Will be conducted by Procurement with assistance of Operation, Engineering or others as required
3. Must be documented by Procurement in a Memorandum to File

INDEX

D

E

F

Quick Order Form

Pourquoi Press

4 easy ways to order:

- **Fax** a copy of this form to 650-348-3425
- **Mail** a copy to P.O. Box 651, San Mateo, CA 94401-651
- **E-mail** your order to PourquoiPress@msn.com (be sure to include all of the information requested below)
- **On line,** place your order at www.BookRegime.com

Please send me ____ copies of *Shock & Awe in Fort Worth*. I understand that I may return them for a full refund, for any reason, no questions asked.

Name ______________________________

Street Address* ______________________________

City, State, Zip Code* ______________________________

Phone ______________ E-mail ______________

Books: _____ copies @ $19.95 each $__________

Sales tax: Add 8.25% for books shipped to California $__________

Shipping to the 48 contiguous states in the U.S.:*

Via UPS Ground (1-6 days), $5.00 for the first book and $1.00 for each additional book $__________

Via UPS 3rd Day Air, $9.00 for the first book and $1.00 for each additional book $__________

Via UPS 2nd Day Air, $12.00 for the first book and $2.00 for each additional book $__________

Shipping to Alaska and Hawaii:*

Via USPS Media Mail (7-10 days), $2.00 for the first book and $1.00 for each additional book $__________

Via UPS 2nd Day Air, $17.00 for the first book and $2.00 for each additional book $__________

Total charges: $__________

Payment method: ☐ Check ☐ Money order ☐ Visa ☐ MasterCard

Credit card number: ______________ Expiration date: ________

Name on card: ______________________________

*At press time, we could not ship to P.O. boxes, military addresses (APO/FPO), or addresses outside the U.S.; however, since this book was printed, our shipping modes have probably expanded (including a distribution point in the UK to minimize international shipping costs), so please check for additional shipping options and destinations at www.BookRegime.com. Shipments to remote locations may take longer than indicated above. All shipping times and rates are subject to change.

Thank you for your order.

Printed in the United States
17663LVS00002B/119

9 780974 994789